IMAGES
of America

THE BENJAMIN FRANKLIN BRIDGE

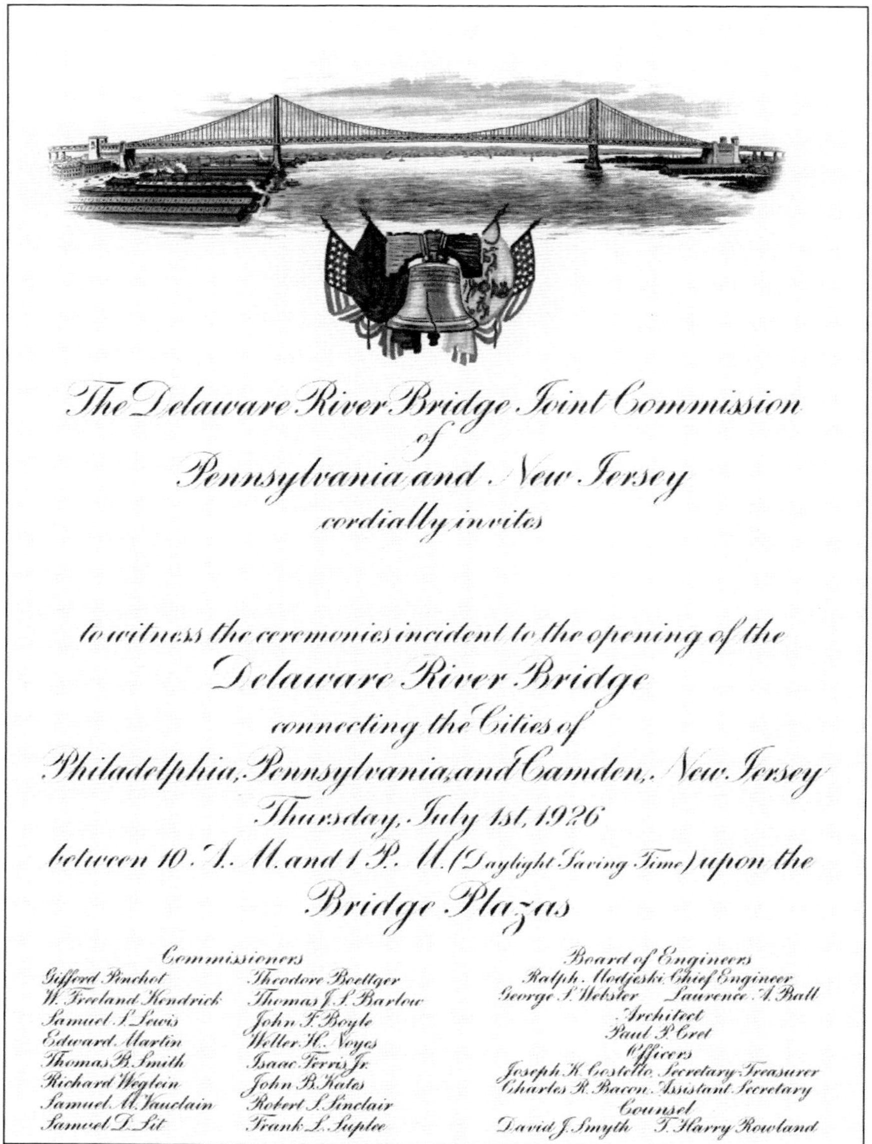

The Delaware River Bridge Joint Commission
of
Pennsylvania and New Jersey
cordially invites

to witness the ceremonies incident to the opening of the
Delaware River Bridge
connecting the Cities of
Philadelphia, Pennsylvania, and Camden, New Jersey
Thursday, July 1st, 1926
between 10 A. M. and 1 P. M. (Daylight Saving Time) upon the
Bridge Plazas

Commissioners
Gifford Pinchot
W. Freeland Kendrick
Samuel S. Lewis
Edward Martin
Thomas B. Smith
Richard Weglein
Samuel M. Vauclain
Samuel D. Lit
Theodore Boettger
Thomas J. S. Barlow
John F. Boyle
Welter H. Noyes
Isaac Ferris Jr.
John B. Kates
Robert S. Sinclair
Frank L. Suplee

Board of Engineers
Ralph Modjeski Chief Engineer
George S. Webster Laurence A. Ball
Architect
Paul P. Cret
Officers
Joseph K. Costello Secretary-Treasurer
Charles R. Bacon Assistant Secretary
Counsel
David J. Smyth T. Harry Rowland

On July 1, 1926, the Delaware River Bridge opened to pedestrians. The invitation shown above from the Delaware River Bridge Joint Commission allowed recipients to attend opening-day ceremonies at the plazas in Philadelphia, Pennsylvania, and Camden, New Jersey. An estimated 100,000 people came to visit and cross what was, at the time, the longest suspension bridge in the world, with a main span of 1,750 feet. The bridge, which cost nearly $37 million, was considered a great feat of engineering. (Authors' collection.)

On the cover: In July 1925, work on the Delaware River Bridge has hit its peak, with a construction high of 1,300 workers. During the summer of this year, no fewer than 10 contracts were underway at various locations on the bridge. Workers in this image are installing the steel curb that will separate the roadway from the never-used track area for trolleys. (Courtesy of the Delaware River Port Authority.)

IMAGES
of America

THE BENJAMIN FRANKLIN BRIDGE

Michael Howard and Maureen Howard

ARCADIA
PUBLISHING

ISBN 978-0-7385-6258-2

Published by Arcadia Publishing
Charleston, South Carolina

Printed in the United States of America

Library of Congress Catalog Card Number: 2008932692

For all general information contact Arcadia Publishing at:
Telephone 843-853-2070
Fax 843-853-0044
E-mail sales@arcadiapublishing.com
For customer service and orders:
Toll-Free 1-888-313-2665

Visit us on the Internet at www.arcadiapublishing.com

To Laura, who was very helpful and patient while her parents wrote this book.

Contents

Acknowledgments

This book would not have been possible without the assistance of many individuals, and we are grateful for their efforts. We thank John J. Matheussen, chief executive officer of the Delaware River Port Authority (DRPA) and president of Port Authority Transit Corporation (PATCO), who graciously allowed us access to historic materials in the company archives, affording us and readers a chance to see remarkable photographs of the bridge, from its inception to the day it opened for business.

We would like to offer particular thanks to a few key individuals from the DRPA for their considerable assistance, including Richard L. Brown; William C. Brooks; Ed Kasuba; Francesca O'Brien; Edward Montgomery; Michael D. Williams; Charles Thompson; Bill O'Connor; Nicholas Settanni, AIA; and Carlton Read.

We also thank Michael F. Britt, from the firm of Modjeski and Masters, and photographer James B. Abbott for their willingness to share unique bridge photographs with us.

Robert Kellman and Dorothy Stanaitis were especially generous in sharing materials from their personal family collections, and we appreciate their assistance.

We are also very grateful to Randi Davis and Nellie McGrew for providing careful editorial review of the manuscript.

The history of the bridge is unique and heavily tied to the region's history as well. Our research was greatly aided by several previous books on bridge history, including *The Building of the Delaware River Bridge Connecting Philadelphia, PA and Camden, NJ,* published just as the bridge opened in 1926 and written by the Delaware River Bridge Joint Commission's assistant engineer Charles Carswell. Another important source for bridge information is *History of the Benjamin Franklin Bridge*, written by Walter S. Andariese and published by the DRPA in 1981. Camden and Philadelphia history provided a backdrop for the bridge's origins and regional impact; please see the bibliography for a complete listing of the sources consulted in exploring this history.

INTRODUCTION

The 1920s were a time of great optimism and prosperity, a decade during which the continued success of America and its cities seemed inevitable. A booming economy and an ever-increasing number of advancements in areas such as radio, transportation, and aviation formed a backdrop for the "Jazz Age" as the 1920s began to truly roar. It seems particularly appropriate that the Benjamin Franklin Bridge (originally named the Delaware River Bridge) rose above Camden and Philadelphia during this heady time.

These two cities had a long-standing historic tie to one another. Quaker William Penn founded Philadelphia, the "City of Brotherly Love," in 1682. Fellow Quaker William Cooper had settled across the Delaware River in 1681, in an area that eventually became Camden. Some years later, Cooper established a ferry between the riverfront area (thereafter known as Cooper's Ferry) and Philadelphia, forging a critical link between the growing settlements.

Philadelphia quickly assumed a prominent role in American history, emerging from its colonial era to become a center of revolutionary activity and the "birthplace of a nation," the place where the Declaration of Independence and the Constitution were drafted. One of the city's favorite citizens, Benjamin Franklin, had helped Philadelphia achieve its eminence as he displayed an astonishing range of talents as a printer, a statesman, and an inventor, and he promoted the new nation's interests as well. By the 1800s, Philadelphia was fast becoming the largest city in the country, with a steady influx of population and the continued annexation of its outlying settlements. In 1876, it was the first American city to host a world's fair, the spectacular Centennial Exposition. (The future chief engineer of the Delaware River Bridge, Ralph Modjeski, reportedly attended the fair during his first visit to America.) The early 1900s were marked by the city's anointing as the "Workshop of the World," an industrial powerhouse that was a center for railroads and many kinds of manufacturing, particularly textiles, printing and publishing, metal products, and leather goods, to name a few. By 1920, Philadelphia's population had reached approximately 1.8 million.

Its neighbor across the river also experienced growth and expansion beyond its earliest days, albeit more slowly. The Cooper's Ferry area developed more like a group of villages or hamlets for some years, finally incorporating in 1828 as the city of Camden (a nod to Charles Pratt, Earl of Camden and Lord Chief Justice of England, who was sympathetic to the colonies in their fight for independence). Growing annoyance with drunken Philadelphian merrymakers who enjoyed Camden's numerous beer gardens and acted as a nuisance to locals played some part in its incorporation as a city, as Camdenites hoped that having their own local government might help to alleviate some of the problems with the carousing visitors. By the 1830s, the Camden and Amboy Railroad came to town, spurring speculation. Throughout the mid-1800s and early 1900s, industries and immigrants began to arrive; Campbell's Soup Company, New York Shipbuilding Corporation, and Victor Talking Machine Company (later the Radio Corporation of America) all located in Camden, along with a significant number of other varied industrial employers. Walt Whitman came to Camden in 1873, residing in the midst of the bustling city until his death in 1892. "On Camden's Supplies the World Relies" was not an empty slogan of this period, and the city's 1920 population reached just over 116,000. As the city's stature grew, its leading businessmen, industrialists, and politicians emerged as a group to promote the concept of a "Greater Camden," boosting the idea of Camden as the increasingly powerful center of Camden County, inextricably connected to its surrounding suburban areas. A centerpiece of this movement was its proponents' efforts to build a bridge between Camden and Philadelphia.

From their earliest days and until the 20th century, Philadelphia and Camden were connected by ferry service only. The first serious proposals to span the Delaware River between these cities had appeared in the 1800s. In 1818, a plan was put forth by Edward Sharp to build a bridge from the area that would become Camden to a then-existent island in the middle of the river; from there, ferry service would have taken passengers to Philadelphia. John C. Trautwine designed a four-span suspension bridge for consideration in 1851; by 1868, committees representing both cities met and issued a report recommending the adoption of a bridge designed by T. S. Speakman. Speakman's concept, a double-draw suspension bridge, received some attention, but plans for it eventually languished, as did all of the proposals from this time. As the two cities experienced their tremendous growth in population and industry, the use of ferries as a primary mode of transport became increasingly inadequate. Civic leaders on both sides of the river saw that the problem of river congestion was worsening, despite the addition of new ferry slips, boats, and facilities. The advent of the automobile also increased pressure to connect the cities by an additional type of crossing, either a bridge or a tunnel.

The efforts that would finally lead to the construction of a bridge between Camden and Philadelphia fully took hold in the 1920s. In 1913, both states began to take serious steps toward securing legislative acts that would allow for a formal commission to investigate bridge or tunnel plans; by 1918, the nascent joint commission advised that a bridge, not a tunnel, should be built. One year later, laws were passed in Pennsylvania and New Jersey that created the Delaware River Bridge Joint Commission (DRBJC), forerunner of today's DRPA. The DRBJC's membership represented both states, and it possessed the "full power and authority to proceed with the construction of the bridge." The commission acted swiftly to appoint a board of engineers to develop recommendations regarding potential bridge sites and final design. In 1921, Congress passed an act that would grant the DRBJC authorization "to construct, maintain and operate" a Philadelphia-Camden bridge, with fixed limits on its location; that same year, the board of engineers suggested that a two-cable suspension bridge should be built from Sixth Street (between Race and Vine Streets, opposite Franklin Square) in Philadelphia to Pearl and Second Streets in Camden. Board members noted that the recommended bridge could be built by July 4, 1926, the nation's sesquicentennial. The DRBJC approved this plan, and work on the bridge commenced in January 1922.

For the next four and a half years, onlookers watched in amazement as the bridge towers and anchorages rose, and then thousands of wire strands were strung back and forth from tower to tower until the cables were formed and banded. After the cables were secured, suspender ropes were hung to support the roadway above the river as citizens of both cities watched it reach out from each tower, meeting at mid-span and each anchorage. When finished, it was the longest suspension bridge in the world, and local residents excitedly awaited its official opening on July 1, 1926, three days ahead of schedule.

On that hot summer day, jubilant crowds listened as bands played and dignitaries from Pennsylvania and New Jersey lauded the bridge's opening. "Isn't it a wonderful thing when your dreams come true?" asked New Jersey governor A. Harry Moore, who went on to remark that "Today Camden and Philadelphia clasp hands." Gifford Pinchot, governor of Pennsylvania, said that "The Delaware River Bridge stands today as a magnificent monument, not only of modern engineering, but of the achieved harmony of our common interests." At 1:00 p.m., the 10- and 12-year-old daughters of Pennsylvania state treasurer Samuel S. Lewis stepped forward at the Philadelphia plaza and released red, white, and blue silken cords, signaling that the Delaware River Bridge was now open to the public. At the same time at the Camden plaza, the four-year-old niece of a New Jersey bridge commissioner did the same. Some 100,000 people were reported to have walked across the bridge roadway that day, until it was closed to foot traffic at 7:00 p.m., in preparation for its opening to vehicular traffic at midnight. Under a headline proclaiming that "Two States Are Linked and Two Neighbor Cities United as Thousands Cheer at Opening of New Bridge," Richard J. Beamish of the *Philadelphia Inquirer* described the then-gray structure as "beautiful as a gossamer web and seemingly as frail" but noted that the bridge did not show any sign of strain as the weight of tens of thousands of people passed over it: "From end to end, the mighty sweep of the superb span was serene, at ease." The sense of wonder evinced by those who crossed the bridge that day can still be felt today and experienced again in the pages that follow.

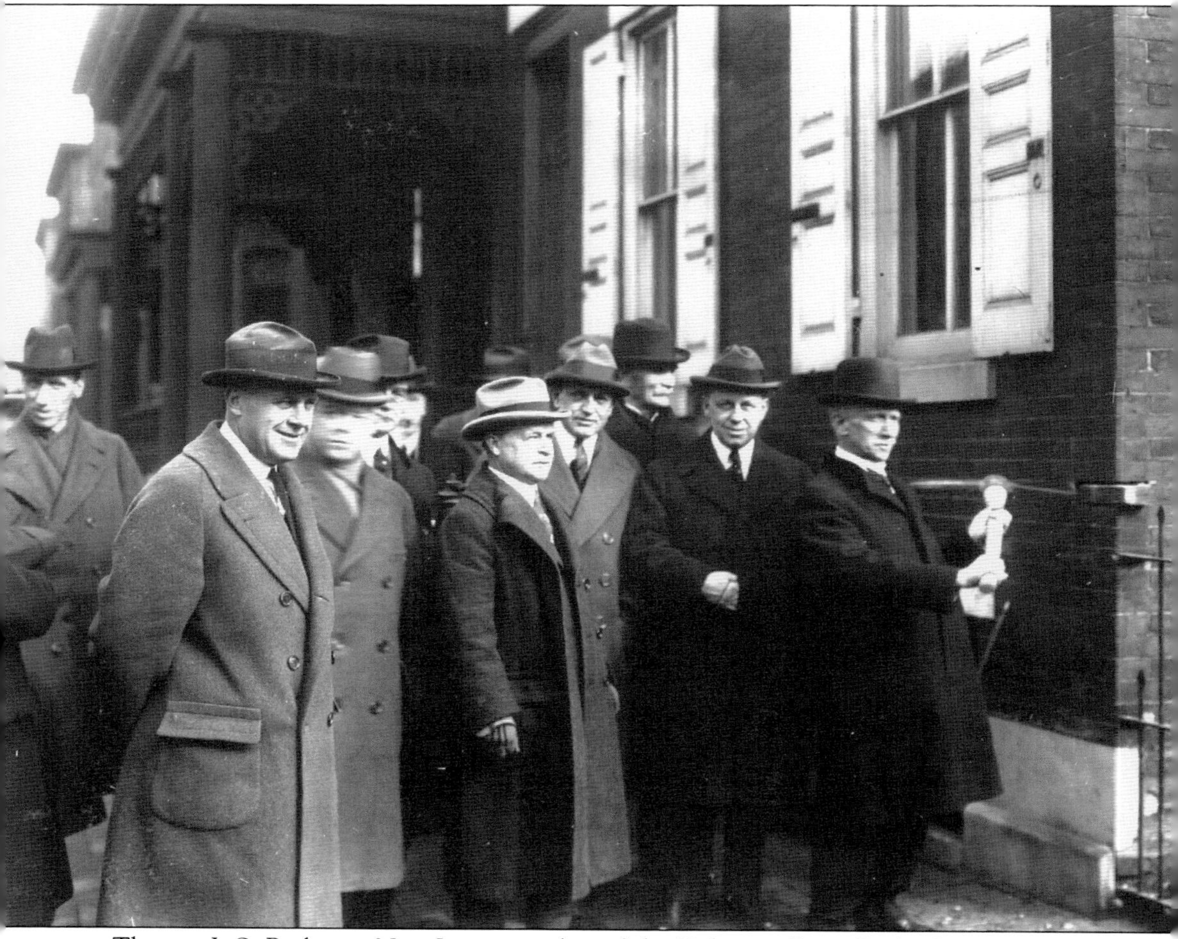

Thomas J. S. Barlow, a New Jersey member of the Delaware River Bridge Joint Commission (DRBJC), is seen here removing the ceremonial "first brick," marking the start of demolition for the bridge's Camden approach.

One

Erected by a Part for the Whole

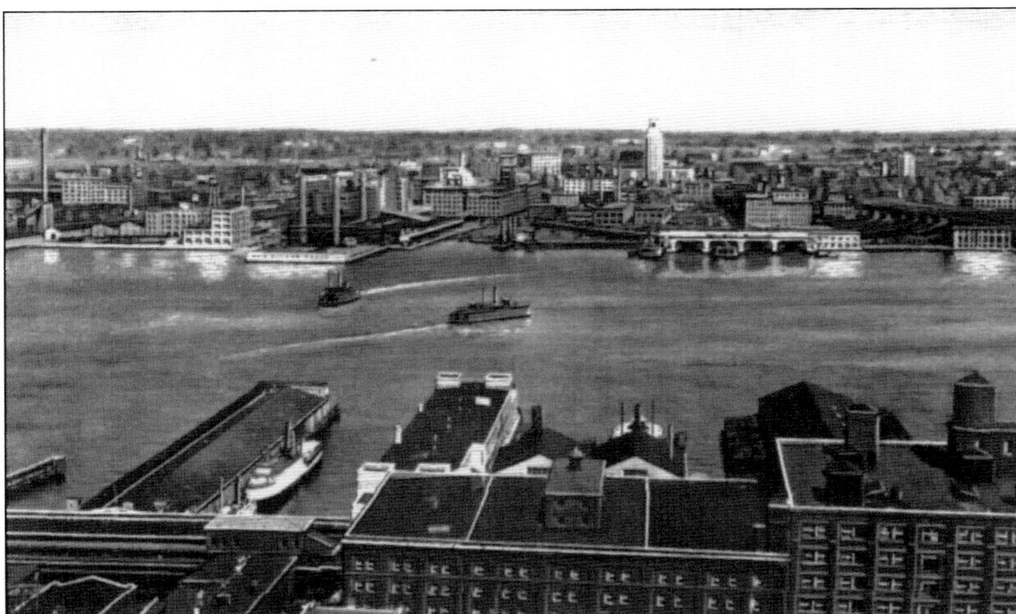

The ferries that long served Camden and Philadelphia are seen in this postcard view of the Camden riverfront. Ferry service was first established in the 1680s, and several ferries eventually operated from various landings between the cities. In October 1926, three months after the Delaware River Bridge opened, the Cooper's Point ferries running between Vine Street in Camden to Vine and Shackamaxon Streets in Philadelphia ceased operation. (Authors' collection.)

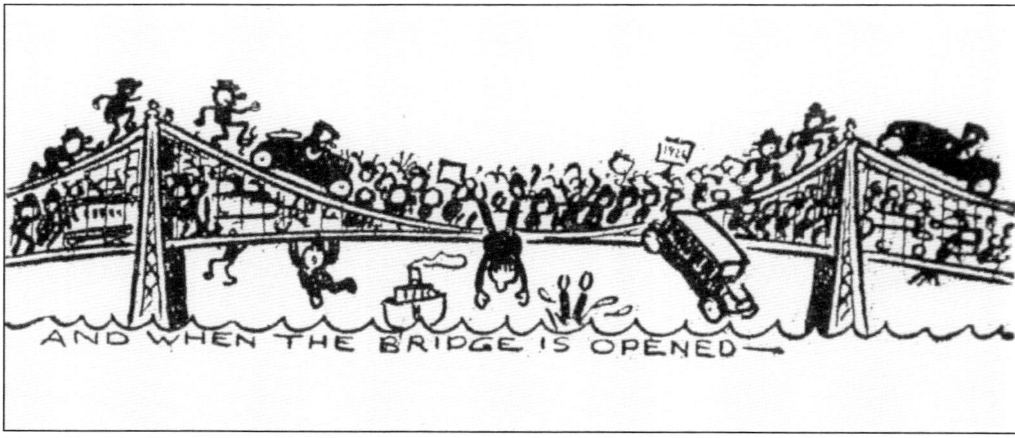

"And When the Bridge Is Opened" is the caption for this whimsical drawing that was published shortly after the bridge was completed. A ferry-type boat is visible beneath the overcrowded bridge; more than two centuries of original ferry service ended in 1952, when the last of the old ferries crossed the Delaware River between Camden and Philadelphia. This image is from Charles Carswell's *The Building of the Delaware River Bridge Connecting Philadelphia, PA and Camden, NJ*.

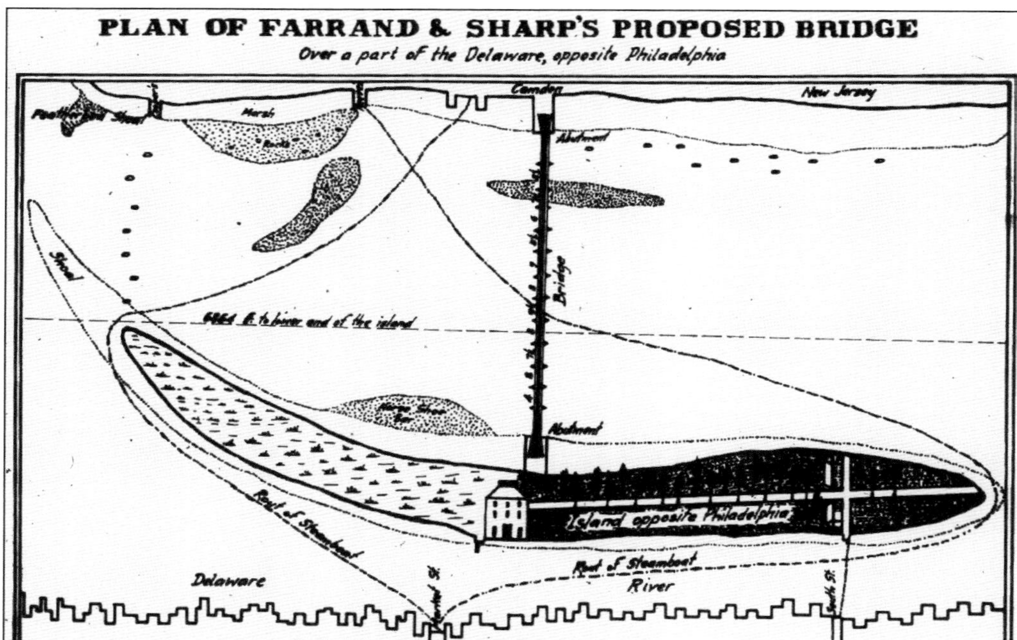

The earliest plan to construct a bridge across the Delaware River between Camden and Philadelphia was proposed in 1818 by Edward Sharp and a Mr. Farrand. Their concept called for building a bridge from the area that became Camden to Smith/Windmill Island; travel between the island and Philadelphia would be made by ferry. Sharp, an early investor in the area, had lots laid out along Bridge Avenue in "Camden Village;" his house still stands at Second and Cooper Streets in Camden.

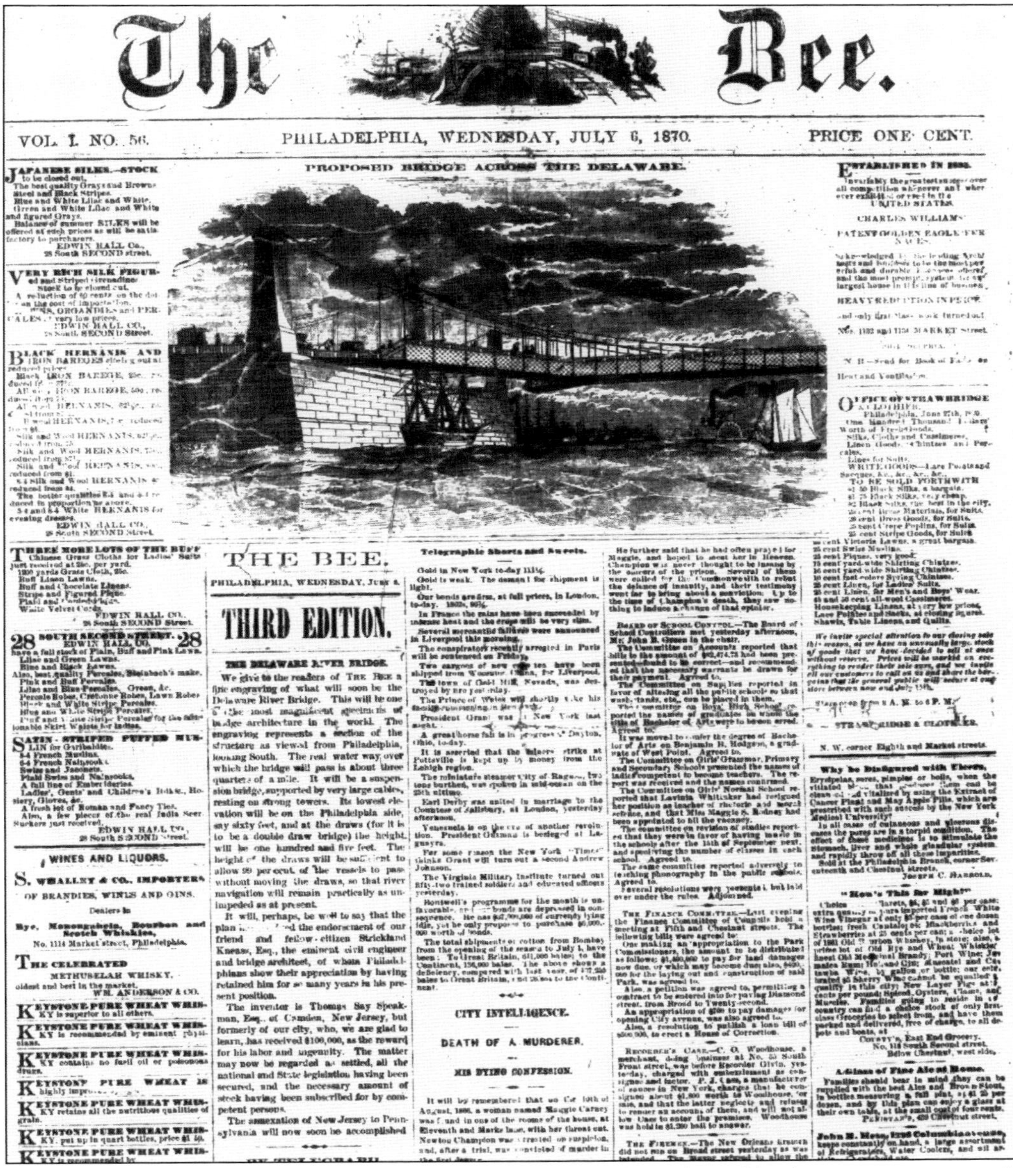

The Bee.

VOL. I. NO. 56. PHILADELPHIA, WEDNESDAY, JULY 6, 1870. PRICE ONE CENT.

PROPOSED BRIDGE ACROSS THE DELAWARE.

THE BEE.

PHILADELPHIA, WEDNESDAY, JULY 6.

THIRD EDITION.

THE DELAWARE RIVER BRIDGE.

We give to the readers of THE BEE a fine engraving of what will soon be the Delaware River Bridge. This will be one of the most magnificent specimens of bridge architecture in the world. The engraving represents a section of the structure as viewed from Philadelphia, looking South. The real water way over which the bridge will pass is about three quarters of a mile. It will be a suspension bridge, supported by very large cables, resting on strong towers. Its lowest elevation will be on the Philadelphia side, say sixty feet, and at the draws (for it is to be a double draw bridge) the height will be one hundred and five feet. The height of the draws will be sufficient to allow 99 per cent. of the vessels to pass without moving the draws, so that river navigation will remain practically as unimpeded as at present.

It will, perhaps, be well to say that the plan [illegible] the endorsement of our friend and fellow-citizen Strickland Kneass, Esq., the eminent civil engineer and bridge architect, of whom Philadelphians show their appreciation by having retained him for so many years in his present position.

The inventor is Thomas Say Speakman, Esq., of Camden, New Jersey, but formerly of our city, who, we are glad to learn, has received $100,000, as the reward for his labor and ingenuity. The matter may now be regarded as settled, all the national and State legislation having been secured, and the necessary amount of stock having been subscribed for by competent persons.

The annexation of New Jersey to Pennsylvania will now soon be accomplished

In 1868, Thomas Speakman proposed the construction of a Y-shaped bridge: a causeway would be constructed from Camden, which, in the middle of the river, would branch off into two diverging roadways with drawbridges leading to Philadelphia. This design attracted some notice, and years later assistant engineer Charles Carswell of the DRBJC quoted its valedictory in his 1926 history of the Delaware River Bridge: "We place above it [Speakman's proposed bridge] no deceptive salutation when we write—'Erected by a part for the whole!'" Although Speakman's bridge was never built, its proponents' exhortation could easily apply to its successor.

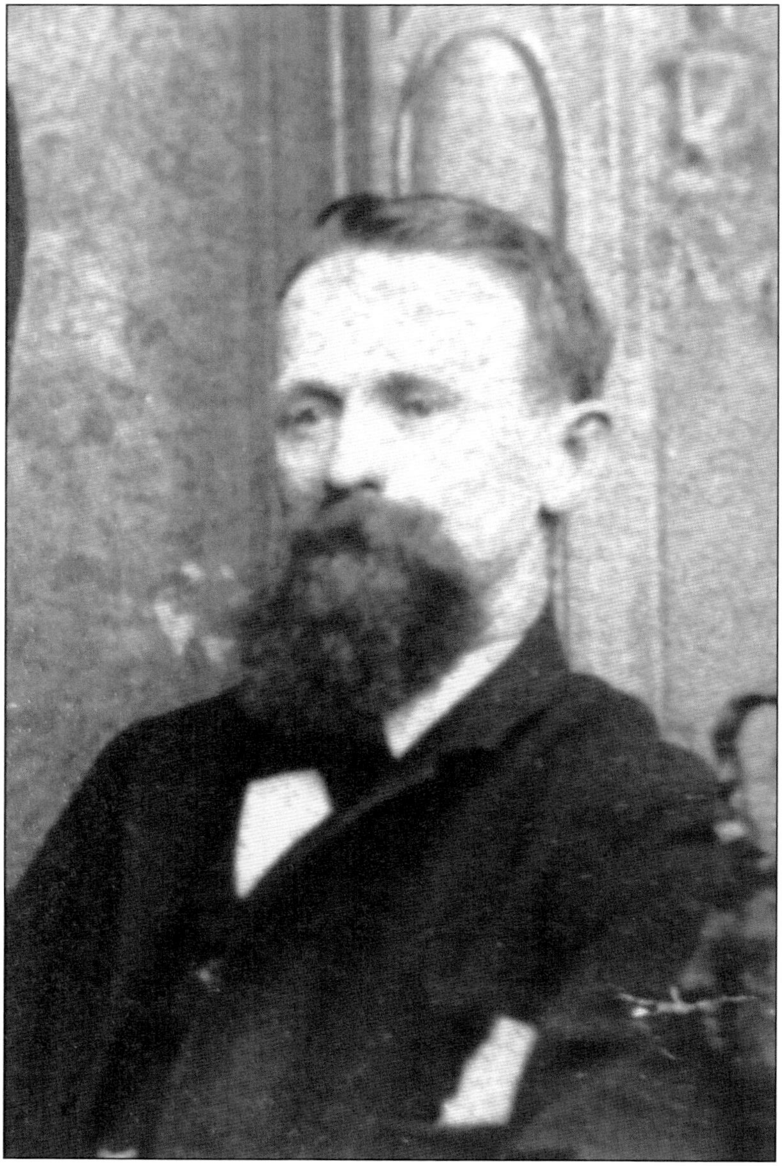

Richard T. Collings, a mayor of Collingswood, New Jersey, and a descendant of one of its founding families, was known as the "Father of the Delaware River Bridge." Collings's efforts helped to bring about laws that permitted preliminary work on bridge planning to begin. As described by Walter S. Andariese in *History of the Benjamin Franklin Bridge*, "these laws stipulated that three or more New Jersey counties could join to initiate a bridge study if they were contiguous and at least one bordered the river to be crossed." Camden and Gloucester county freeholders were supportive, but Burlington County officials voted down Collings's request for appropriations. As he later sat commiserating with local politicians at a Mt. Holly hotel, Collings overheard a nearby group of farmers complaining of their frustrations with crossing the river by ferry and recruited them to tell their stories to the reluctant freeholders, who then reversed their earlier decision. An early member of the DRBJC, Collings died in 1920, never seeing the realization of his efforts. (Courtesy of the Collingswood Public Library.)

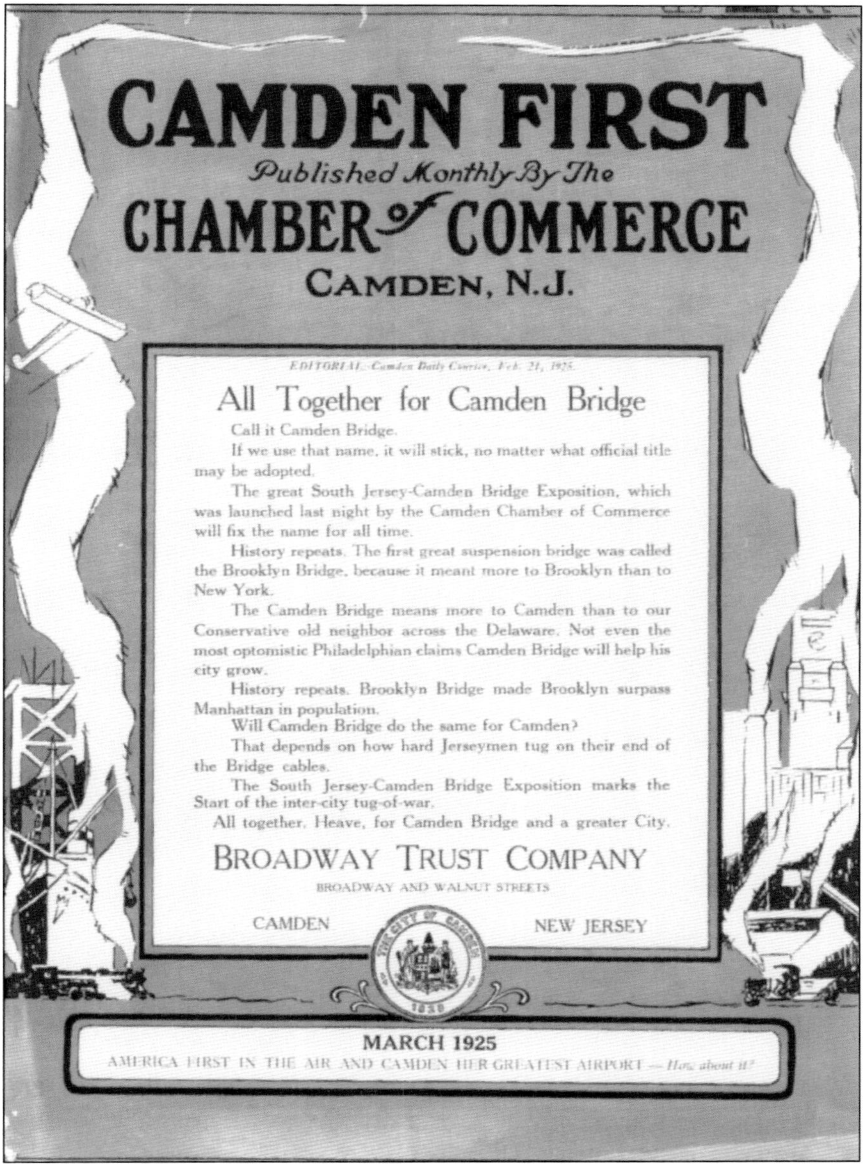

This is the cover of *Camden First* from March 1925, with an editorial about the Delaware River Bridge then under construction, advising readers to "Call it Camden Bridge." The pages of this monthly publication contained numerous articles advancing the "Greater Camden" concept and extolling the seemingly boundless opportunities that the opening of the bridge would bring to Camden. Newspaper editorials of the day and *Camden First* writings reveal their authors' conviction that the bridge would finally allow Camden to reach its fullest potential, with an expected boom in population and industry to follow. This expectation led Camden bridge proponents to propose that their city would eventually reach a status closer to its much larger and more famous "friendly rival" across the river. The cover puts it succinctly: "The Camden Bridge means more to Camden than to our Conservative old neighbor across the Delaware . . . History repeats. Brooklyn Bridge made Brooklyn surpass Manhattan in population. Will Camden Bridge do the same for Camden?" (Courtesy of Camden County Historical Society.)

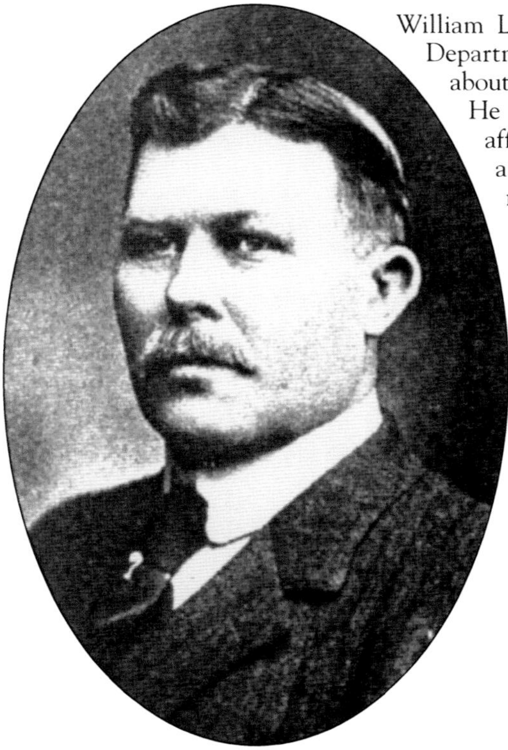

William Leonard Hurley was the owner of the Hurley Department Store in Camden, with other stores scattered about New Jersey, Pennsylvania, and Delaware. He was active in Camden City civic and business affairs and a bridge promoter. As plans to build a bridge neared realization, Hurley put forth his recommendation that it should be situated from Washington Square in Philadelphia to Berkley and Fifth Streets in Camden. (Courtesy of Camden County Historical Society.)

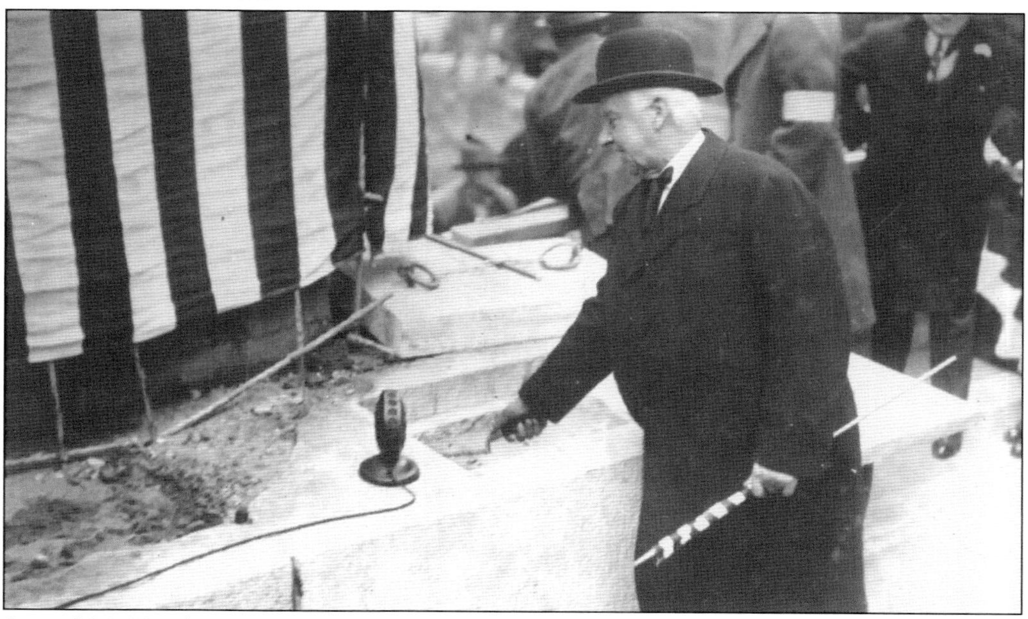

Samuel M. Vauclain, president of Baldwin Locomotive Works, is seen here participating in a ceremony that took place during the laying of the bridge cornerstone on December 12, 1924. Vauclain, another bridge booster and a Pennsylvania member of the DRBJC, thought that it should be built from Spring Garden Street in Philadelphia to Main Street in Camden, and he gave presentations espousing this proposal in both cities.

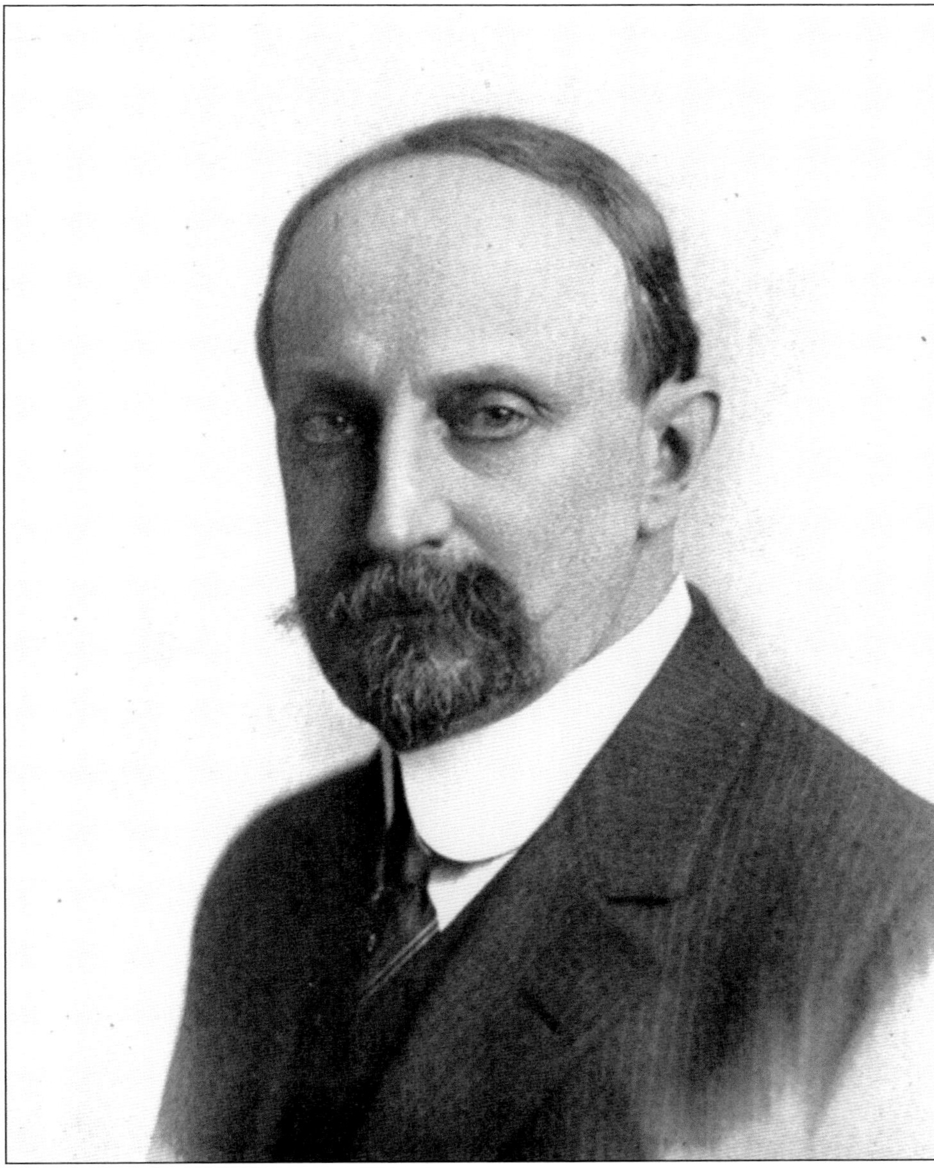

Ralph Modjeski was the chief engineer of the Delaware River Bridge. Born in Poland, Modjeski studied in Paris and started practicing as a consulting bridge engineer in the early 1890s in Chicago. Before long, he established a successful career, and some of his well-known bridges from this era include spans over the Columbia and Willamette Rivers in Oregon; a seven-span railroad and highway bridge across the Mississippi at Rock Island, Illinois; and the McKinley Bridge at St. Louis, Missouri. Modjeski was admired by engineering peers and bridge workers alike. As work progressed on the Delaware River Bridge, he visited the site often and did not hesitate to experience its working conditions personally, whether descending into the foundation caissons below the river or walking across its temporary footwalks, high in the air. At the bridge's opening, Philadelphia and Camden hailed the achievement of the engineer who had helmed the construction of what was then the world's longest suspension bridge. This image is from Charles Carswell's *The Building of the Delaware River Bridge Connecting Philadelphia, PA and Camden, NJ.*

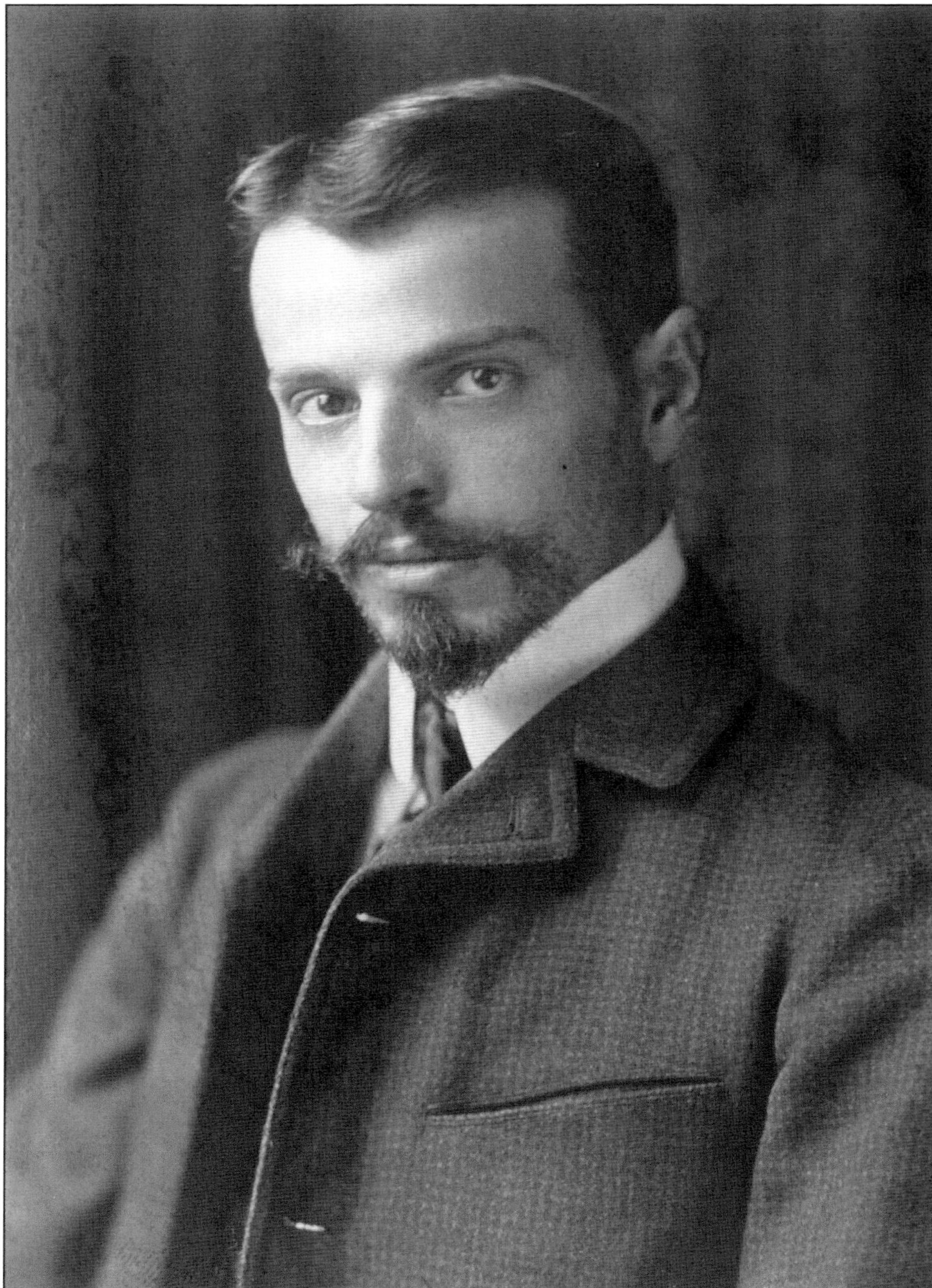

Paul Cret, the bridge's architect, studied at the Ecole des Beaux Arts in France and was a professor of design at the School of Architecture of the University of Pennsylvania. As architect, his task was to give the bridge its overall look, in keeping with Modjeski's engineering specifications. His graceful work is particularly evident in the design of the anchorages and the original appearance of the bridge's plazas. A further example of his work is the Rodin Museum in Philadelphia. (Collections of the University of Pennsylvania Archives.)

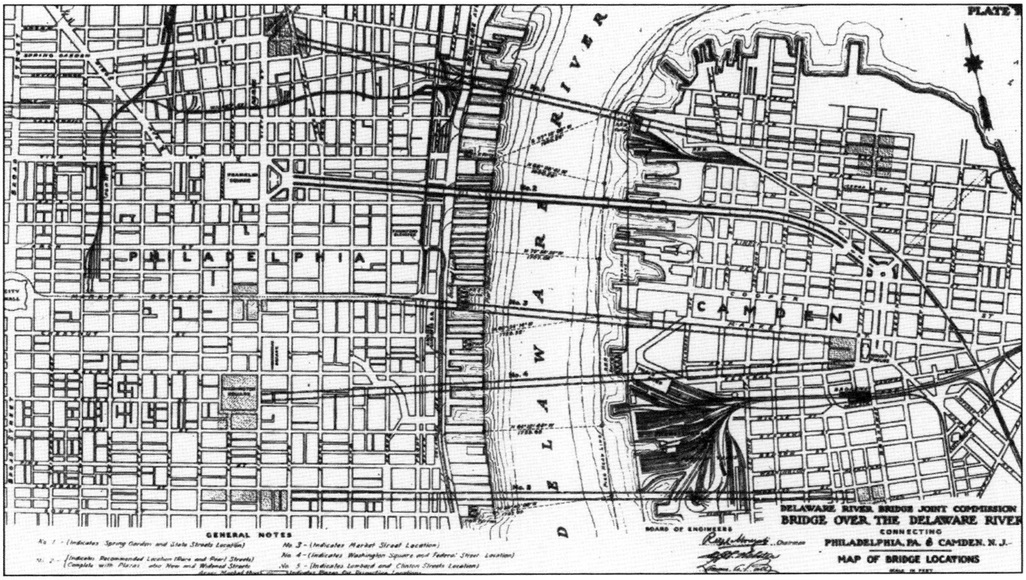

On February 15, 1921, the DRBJC was formally tasked to build a bridge between Philadelphia and Camden. As seen in the plan above, five locations were identified as potential bridge sites. The five locations were sited between Fifth/Spring Garden Streets and Fifth/Lombard Streets in Philadelphia and Sixth/State Streets and Fifth/Berkley Streets in Camden. Land and river borings taken at each selected area were used to determine the best and most economical site to build the bridge.

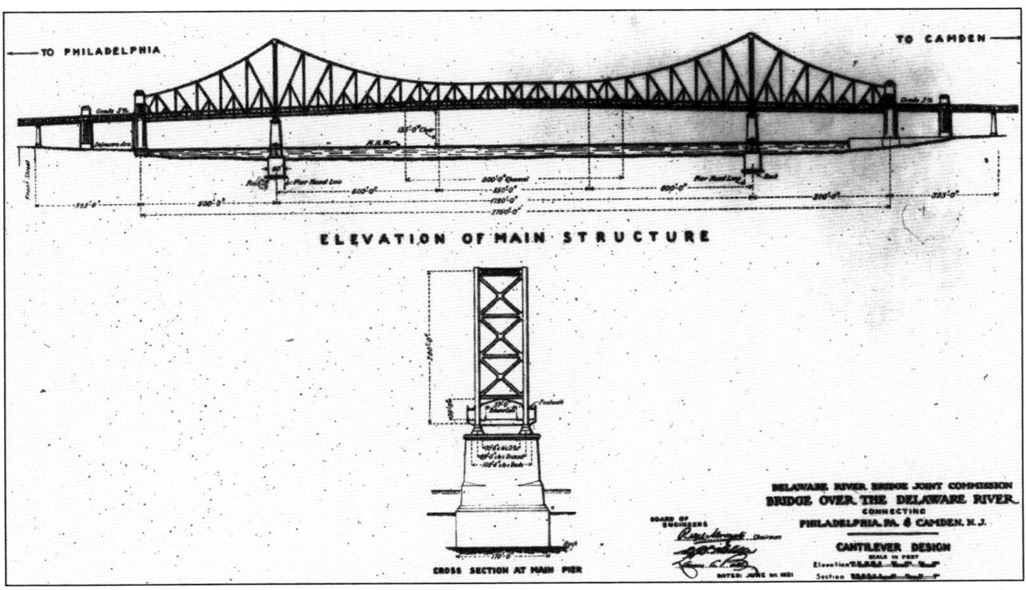

In addition to determining the final bridge site, the board of engineers was asked to evaluate the type of bridge to be constructed. This drawing from 1921 shows a cantilever-designed bridge, one of the bridge styles evaluated during the planning phase.

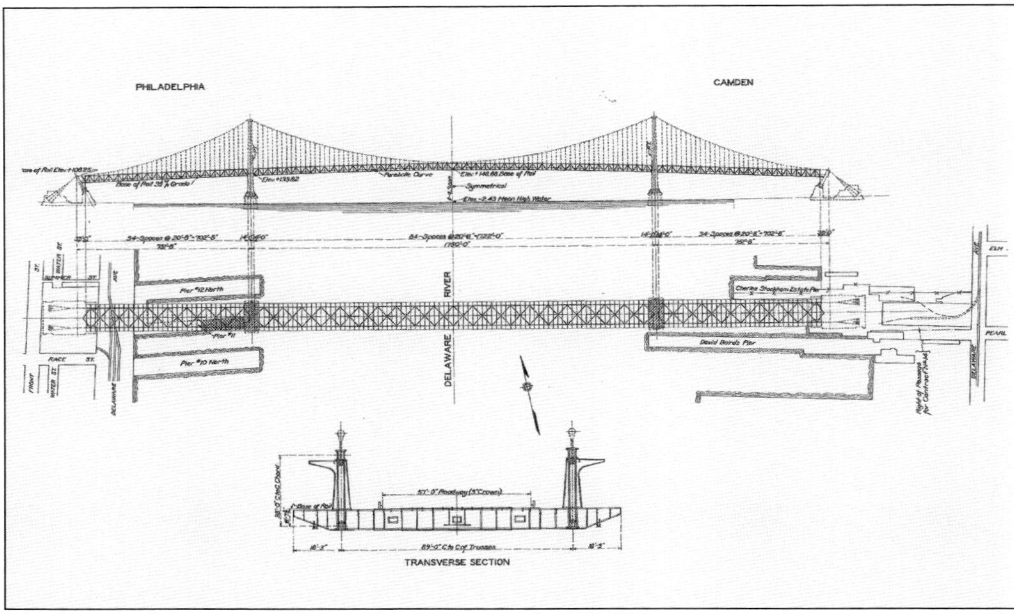

The board of engineers submitted a report to the DRBJC recommending construction of a suspension bridge between Sixth and Race Streets in Philadelphia to Pearl and Second Streets in Camden, curving to Penn Street between Sixth and Seventh Streets. The bridge would have a main span of 1,750 feet and a clearance above the river of 135 feet, dimensions that were dictated by the Department of War.

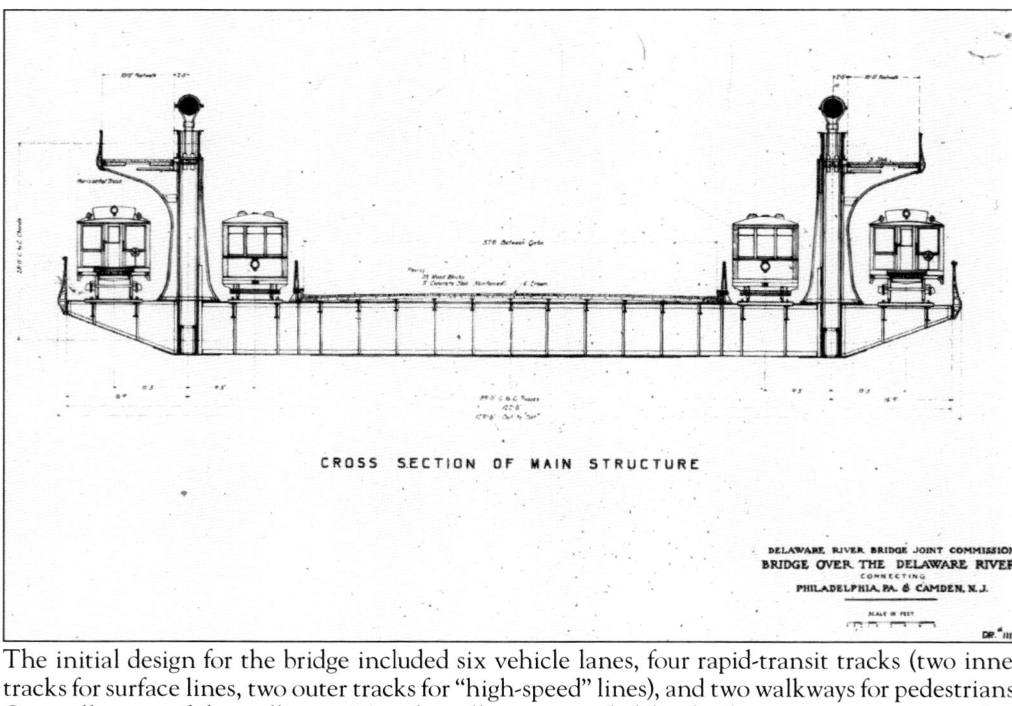

The initial design for the bridge included six vehicle lanes, four rapid-transit tracks (two inner tracks for surface lines, two outer tracks for "high-speed" lines), and two walkways for pedestrians. Carswell wrote of the walkways, "Ample walks are provided for the foot passengers, rare as they may become in a completely motorized age."

During design, sketches were prepared to reflect the proposed aesthetic look of the bridge. This image is a rendering of the bridge's suspended spans. The drawing reflects the multiple modes of travel incorporated into the design of the structure: vehicular, rapid transit, and pedestrian.

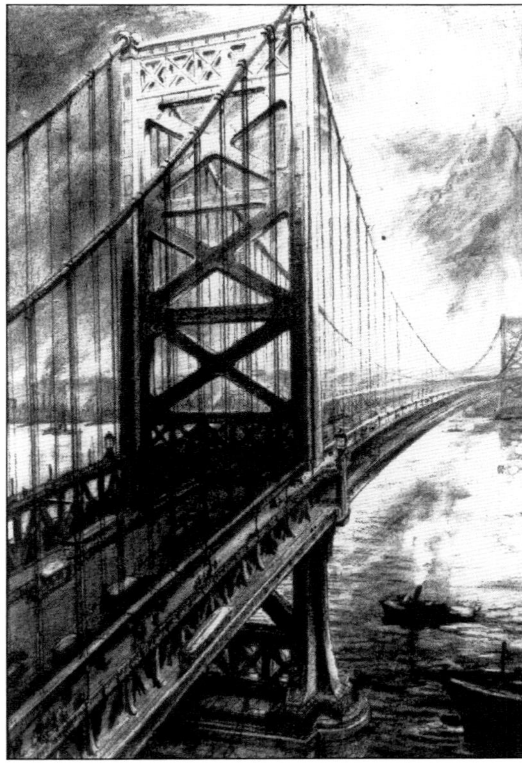

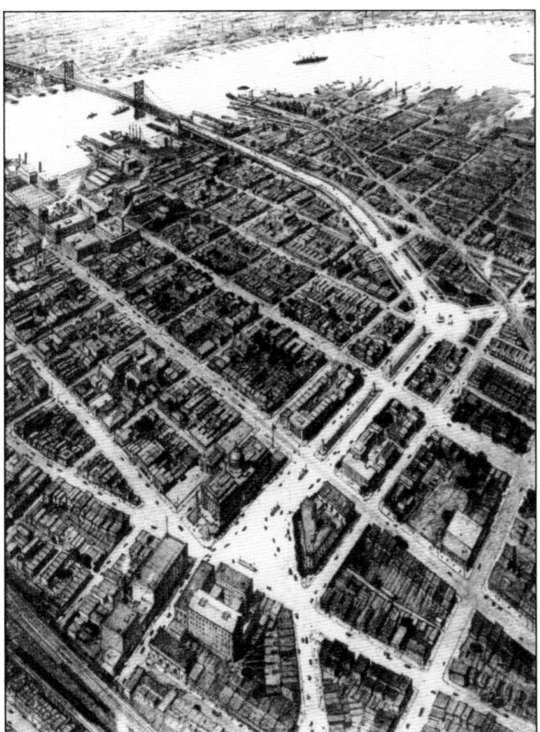

Paul Cret's intent for the Camden approach was to create a "grand avenue" leading to the bridge. A plaza consisting of a full city block was envisioned for the Camden terminus of the bridge, and Camden's main artery and bridge entrance, Broadway, would be extended two blocks past Market Street to the new plaza. Broadway would also be widened between Federal Street and the plaza. Monuments were planned at each intersection to create a gateway effect.

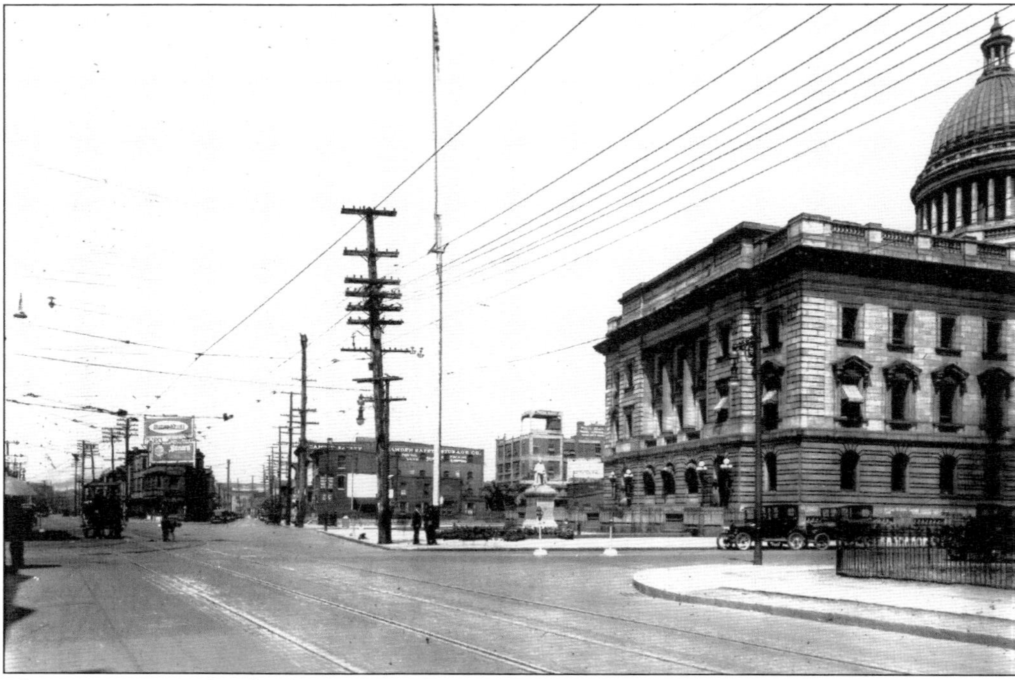

The intersection of Federal Street, Arch Street, and Broadway in Camden, as planned by Paul Cret, would have been the southern end of the "grand avenue" leading to the bridge. The building on the right side of the image is the Camden County Courthouse, which was razed in the 1950s to make way for Lit Brothers department store.

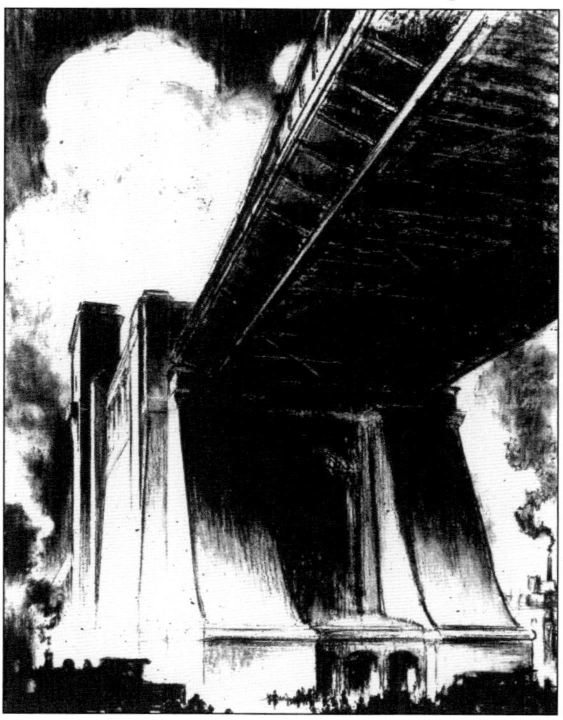

This drawing reflects the monumental scale of the proposed bridge anchorages. A massive structure is required to resist the pull of the bridge's cable due to loads associated with a suspension bridge. The anchorage was also to serve as a station, which was never used, for pedestrians wanting to transfer from street level to the walkway or to the rapid-transit lines. Elevators within the anchorage towers would have moved people to the various levels within the anchorage.

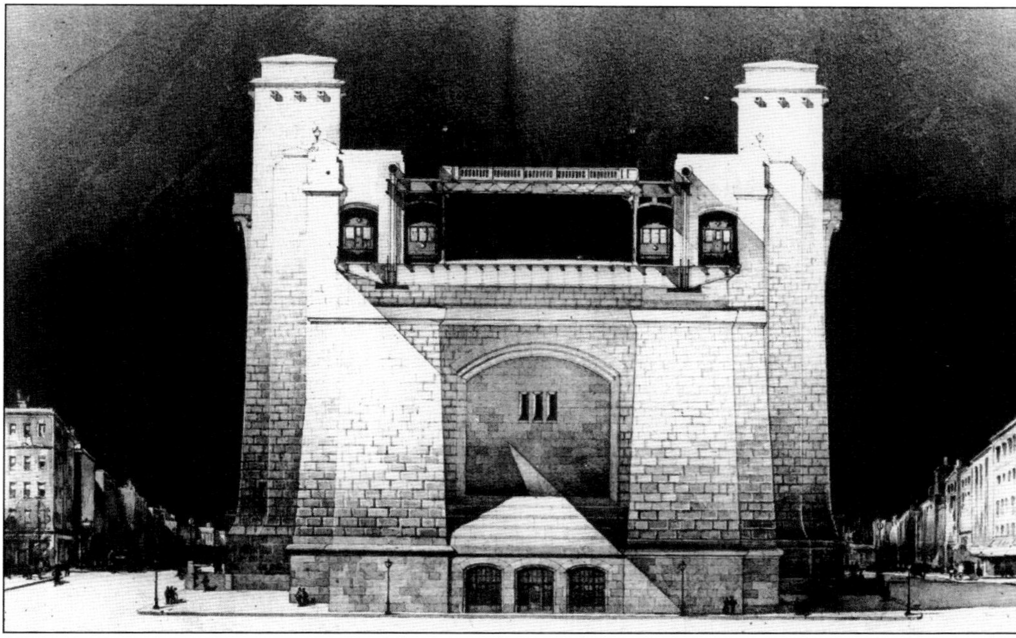

This elevation view by Cret reveals the relationship between the Philadelphia anchorage and the surrounding community. The planned station within the anchorage would have served a large industrial, commercial, and residential population. Residents and customers could have used the anchorage station for access to the rapid-transit lines across the bridge or the Philadelphia Transit Company's Frankford-Elevated Railroad along Front Street.

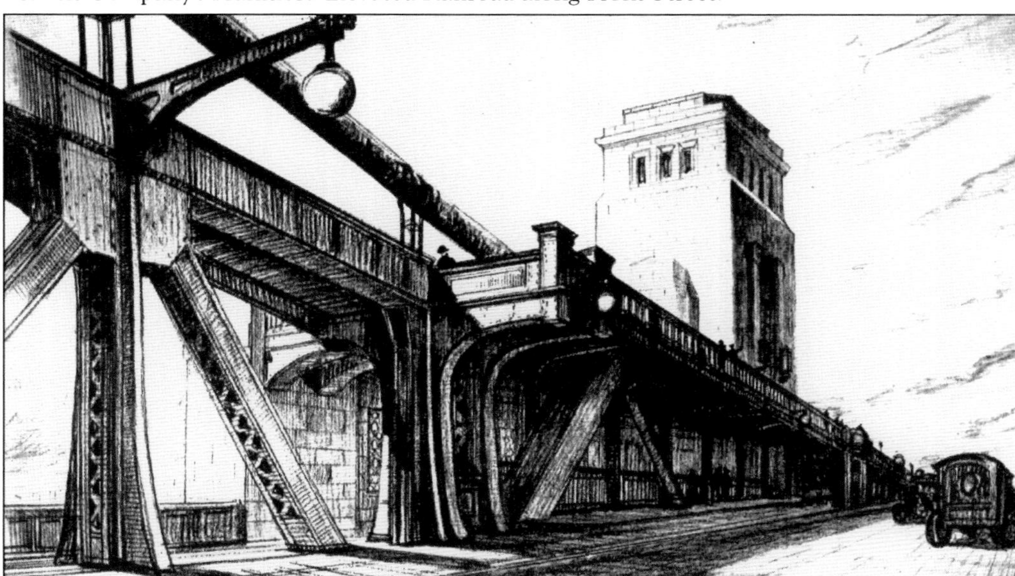

This drawing depicts the approach roadways and the side suspension span. The towers at each anchorage contained elevators that would have transported people from the street level to the walkway or rapid-transit station levels. At the walkway level at each anchorage, pedestrians would have entered a large elevator lobby with decorative tiles commemorating milestones of transportation in the United States.

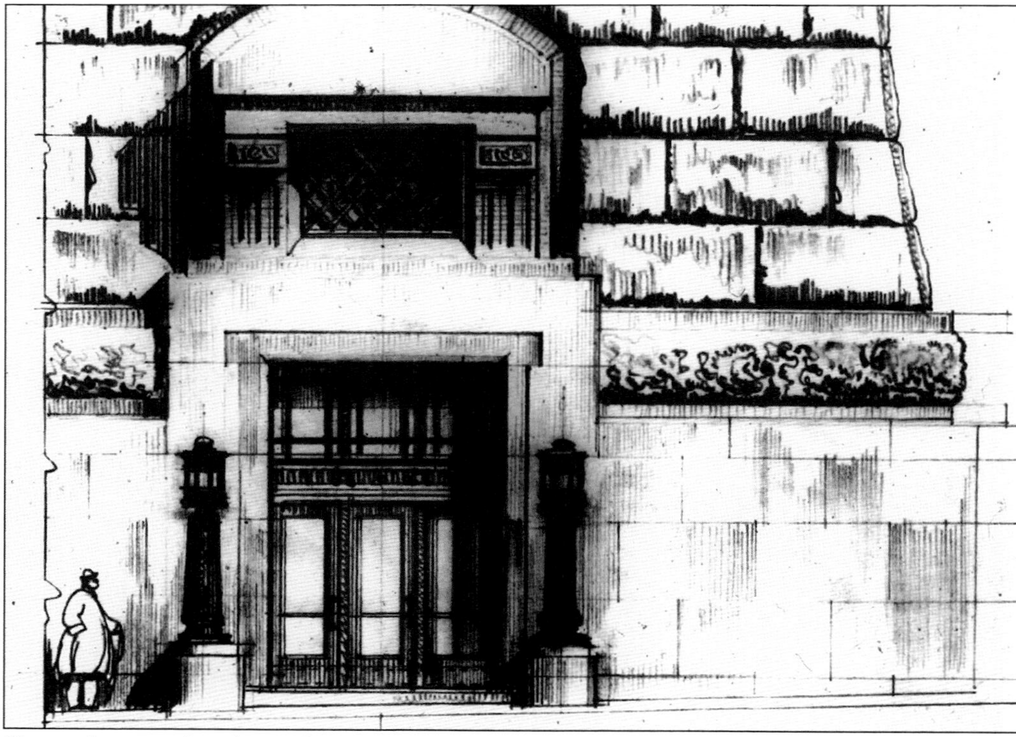

Shown here is the planned street-level entrance into the anchorage. Beyond this entrance, pedestrians would have entered the elevator lobby. These elevators were to provide access to the various levels inside the anchorage. Although the lobbies and elevators were constructed, they were never opened to public use.

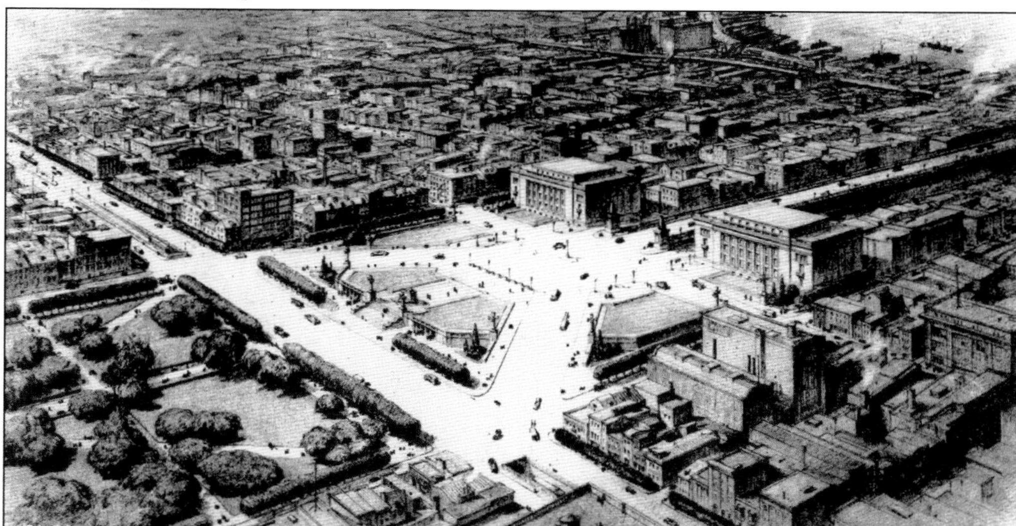

Similar to the "grand avenue" concept for Camden, the Philadelphia terminus of the bridge featured a "gateway" design. Although the plaza's trapezoidal shape and the tunnel beneath Sixth Street have been modified or relocated, the general concept seen here was retained when the bridge opened.

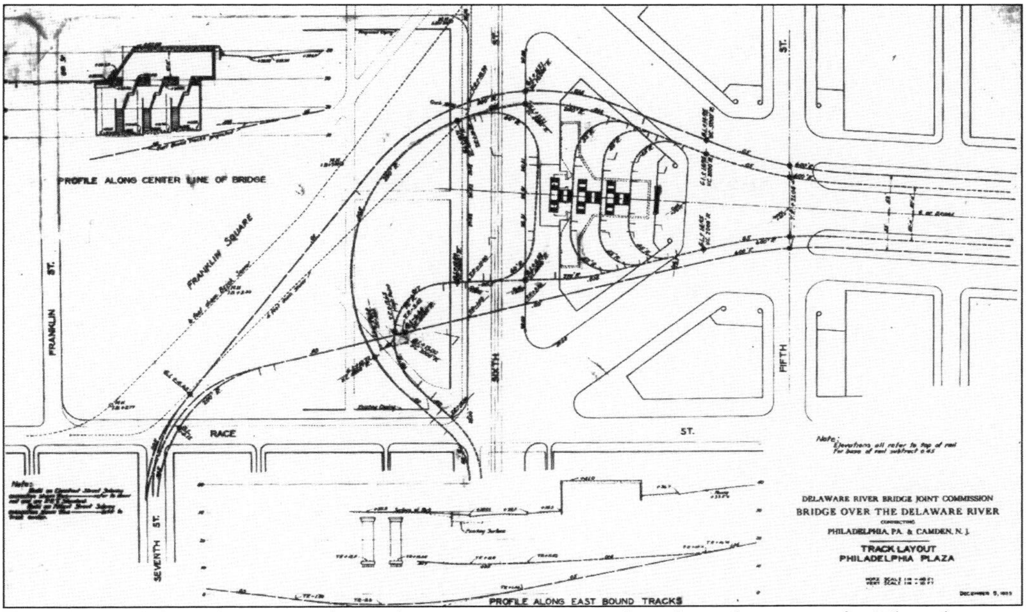

The Philadelphia plaza was planned to meet operational and aesthetic needs. This drawing shows how the bridge's transit lines would descend into an underground station beneath the plaza. A provision would allow the rapid-transit surface cars (or trolleys) to enter a four-track station with platforms or bypass the station and enter the Philadelphia transit system. Tracks for the high-speed trains would bypass the station and turn south on to Seventh Street.

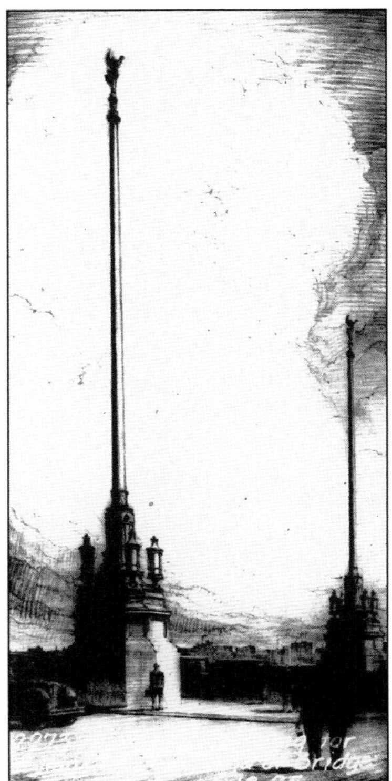

This artist's rendering portrays a pair of pylons to be erected at each end of the bridge. Each steel mast would be topped by "Winged Victory" figures 75 feet above the roadway, to watch over the bridge and its traffic. The eight-foot-high bronze figures, which came to be referred to as "angels," stood at each end of the bridge until 1949, when they were removed to accommodate roadway widening from six to eight lanes.

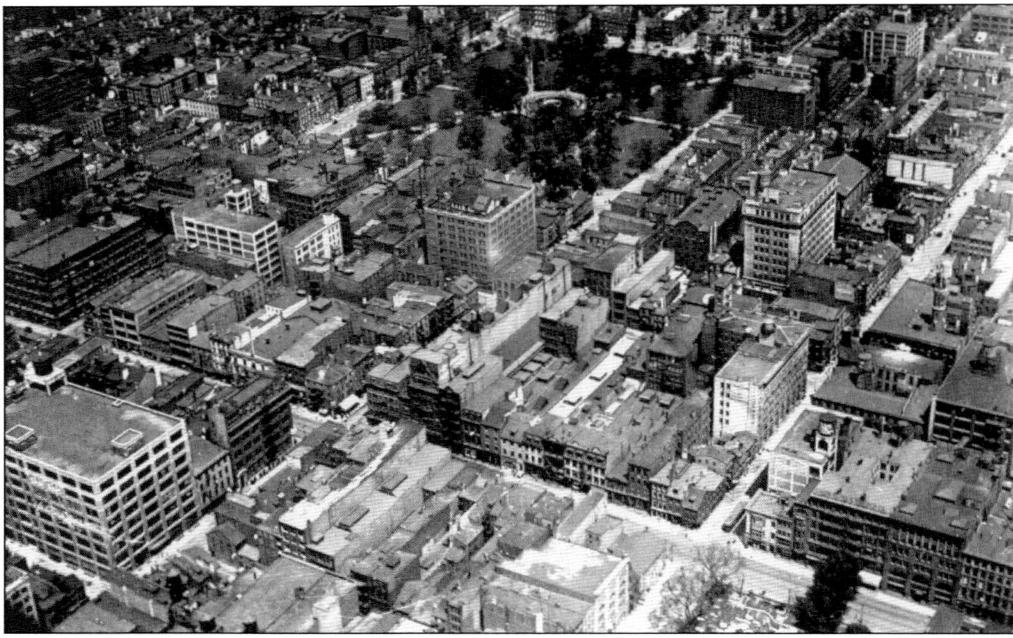

The aerial photograph shows the area surrounding Franklin Square in Philadelphia, prior to bridge construction. One reason that this site (Sixth Street between Race and Vine Streets) was chosen was because of the expectation that it would provide the most favorable traffic conditions compared to the cost of constructing the bridge. Those favorable conditions would result in heavier-than-anticipated traffic volumes, which would require Philadelphia to eventually modify the approach streets.

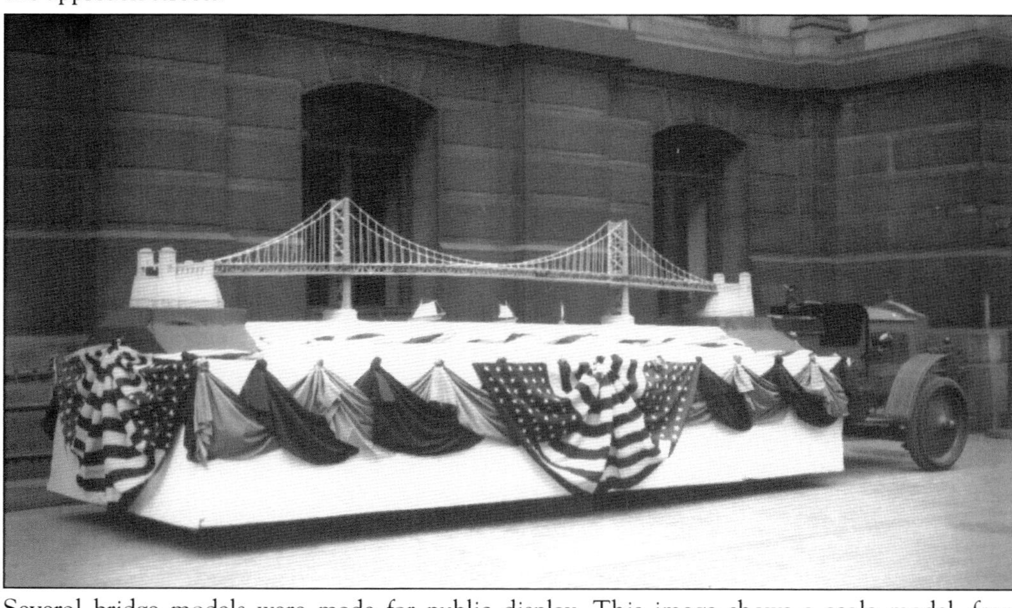

Several bridge models were made for public display. This image shows a scale model, from anchorage to anchorage, on a platform pulled by a truck parked in front of Philadelphia City Hall. This float was part of a parade to celebrate the start of bridge construction. (Courtesy of Modjeski and Masters, Inc., Mechanicsburg, PA.)

Leon Hermant, an internationally known sculptor, is seen here developing the plaster model for the four "Winged Victory" bridge angels. Hermant also sculpted coats of arms representing Pennsylvania, New Jersey, Camden, and Philadelphia for the north and south faces of both anchorages. The plaster model of the New Jersey coat of arms can be seen in the background.

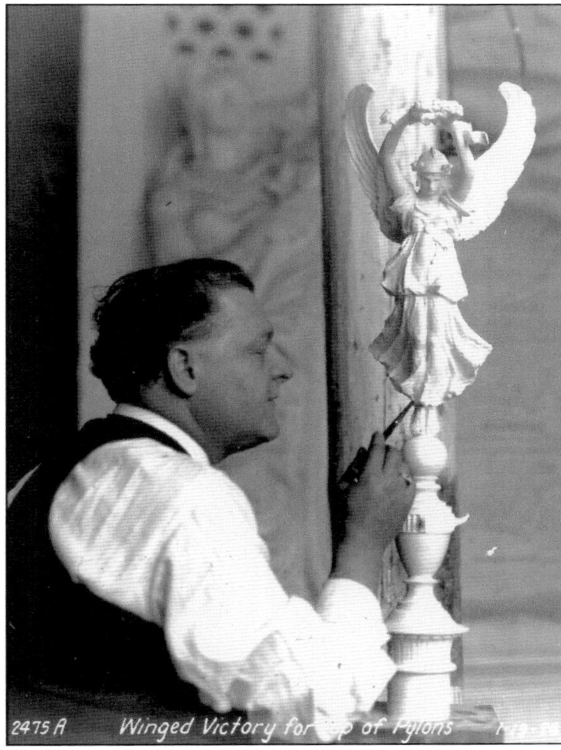

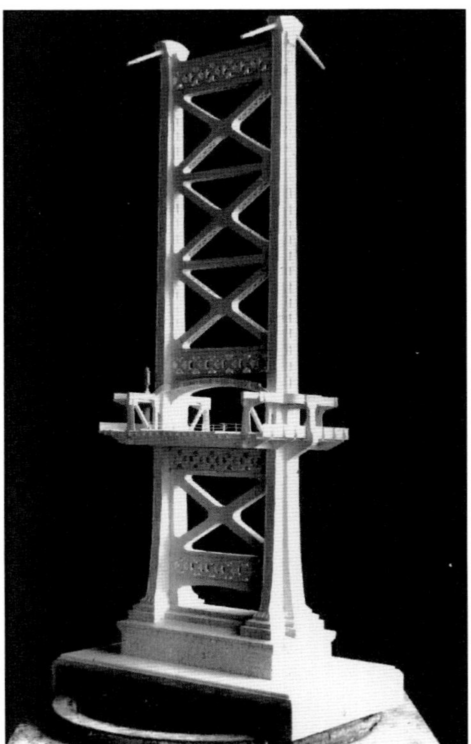

This image shows a three-quarter view of the bridge tower. The model depicts the architectural detailing that will be associated with the actual tower, such as cross-bracing, decorative lighting, pedestrian walkways, rapid-transit tracks, and even the individual granite blocks along the exterior face of the tower pier.

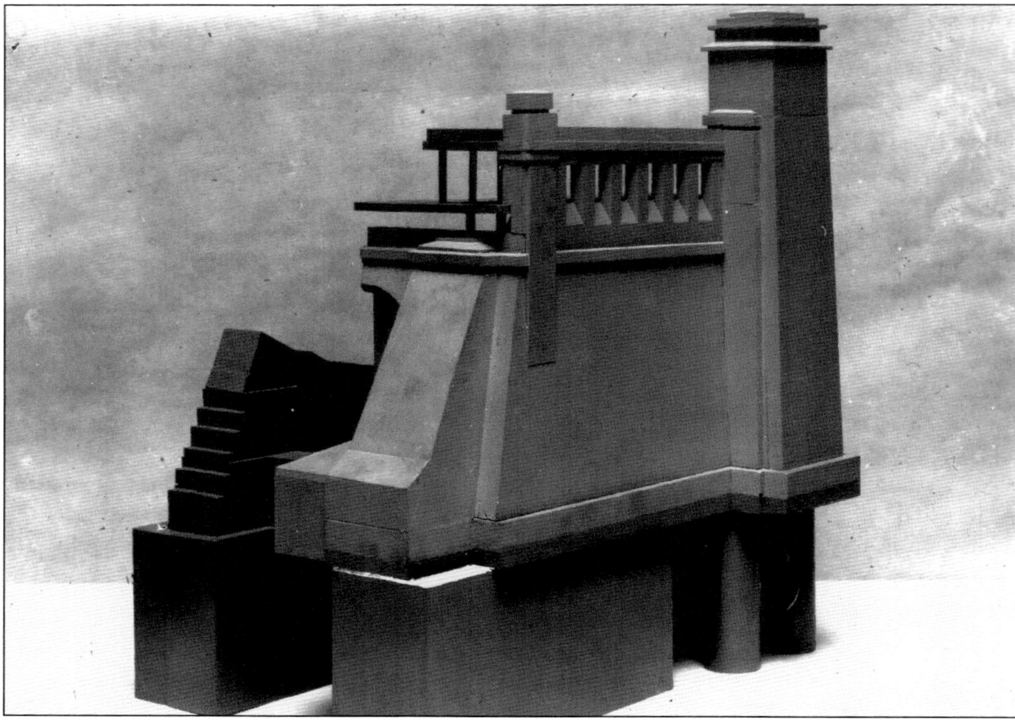

Seen here is a three-quarter view of the bridge anchorage. Unlike the tower model shown in the preceding image, this model also includes the caisson foundation. The anchorage foundation consists of two rectangular and eight circular concrete blocks. Each of the footings was to be founded on bedrock, providing a solid base to support the anchorage.

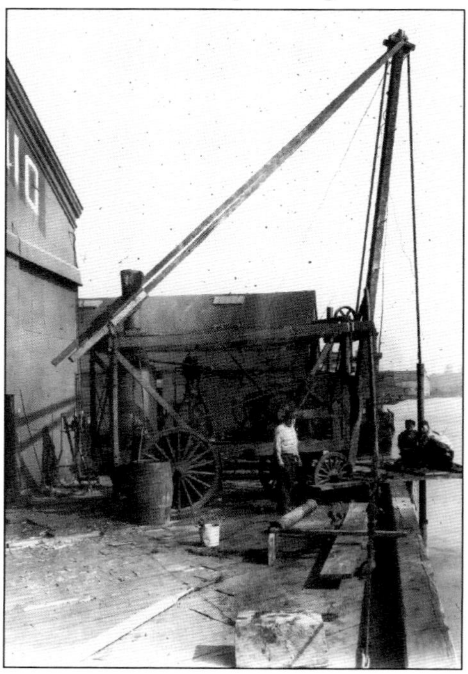

As part of recommending a final bridge site, the board of engineers had a contractor take river and land borings along each proposed location to identify the type of soil and depth of bedrock. This photograph from 1921 shows river borings being taken at one of the proposed bridge tower sites.

Two

Spanning the Delaware

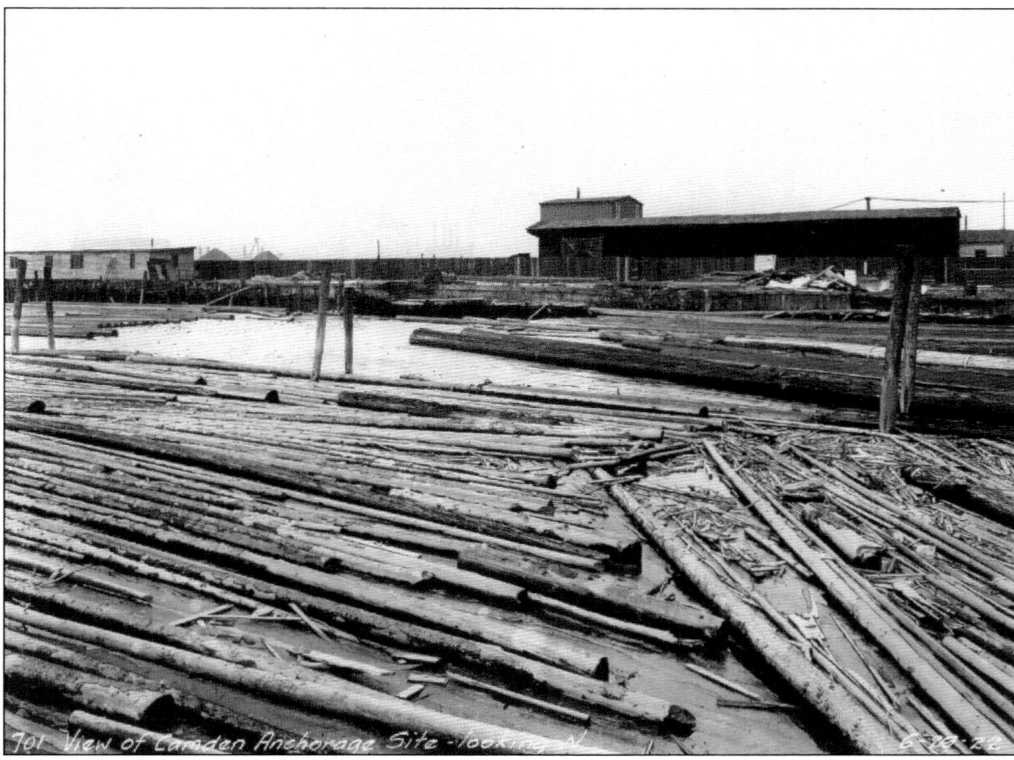

This image, just north of the David Baird Company lumberyard in Camden, shows the future location of the anchorage. Before construction, this area was used by Baird's company for the storage of logs along the riverbank.

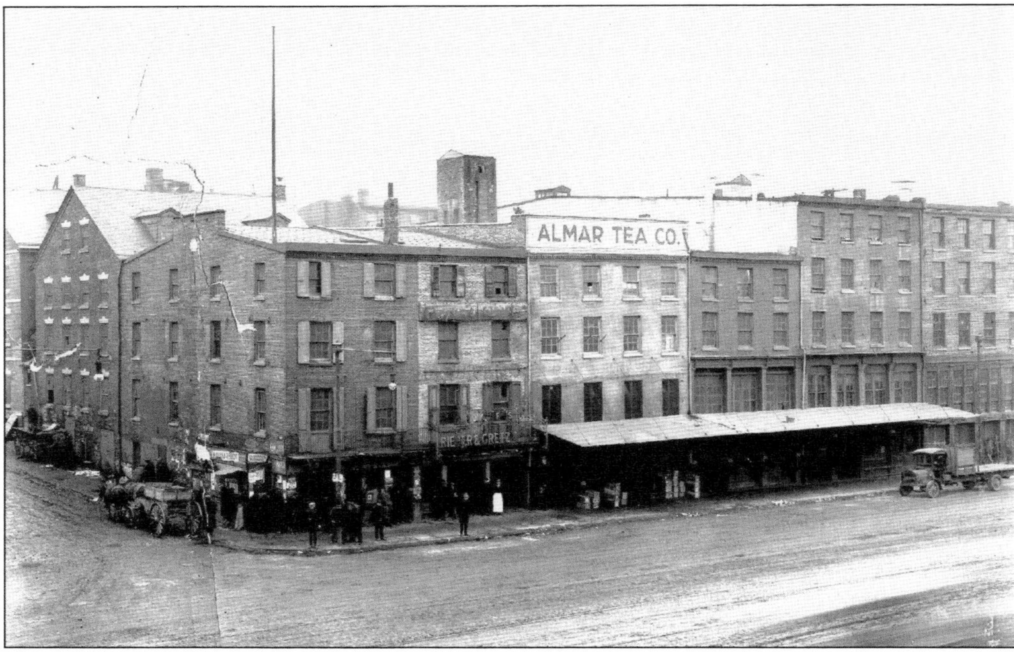

Before 1922, the northwest corner of Race Street and Delaware Avenue in Philadelphia comprised several commercial properties. Demolition of this group of buildings began in April 1922 for the anchorage foundation excavation.

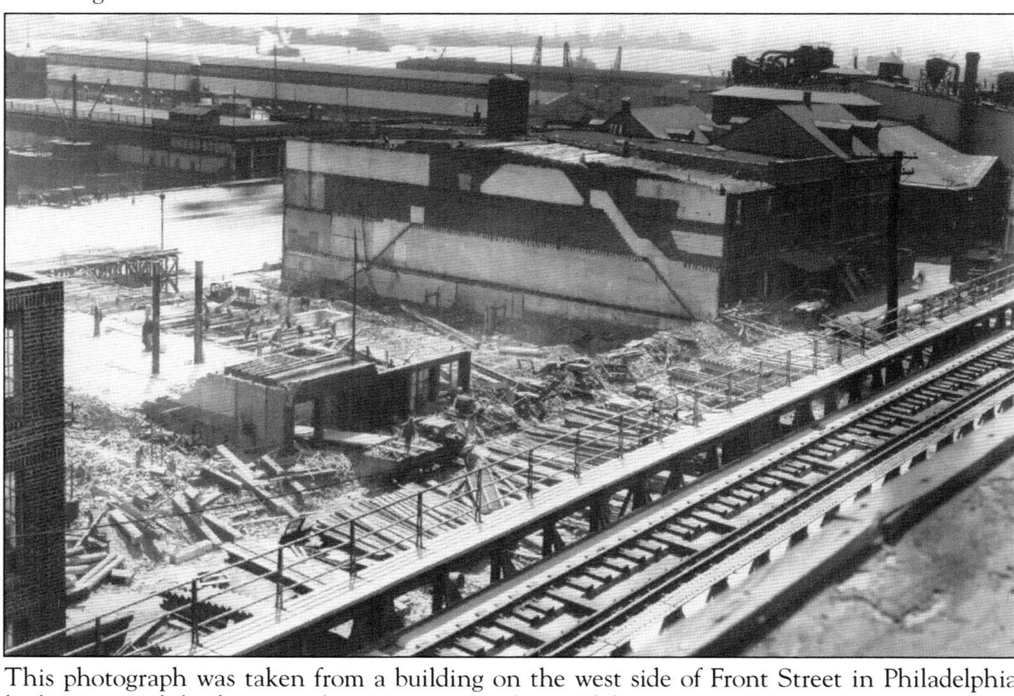

This photograph was taken from a building on the west side of Front Street in Philadelphia, looking toward the future anchorage site. Demolition of the existing structures is well underway. The close proximity between the elevated structure and the anchorage site is evident by the tracks in the foreground.

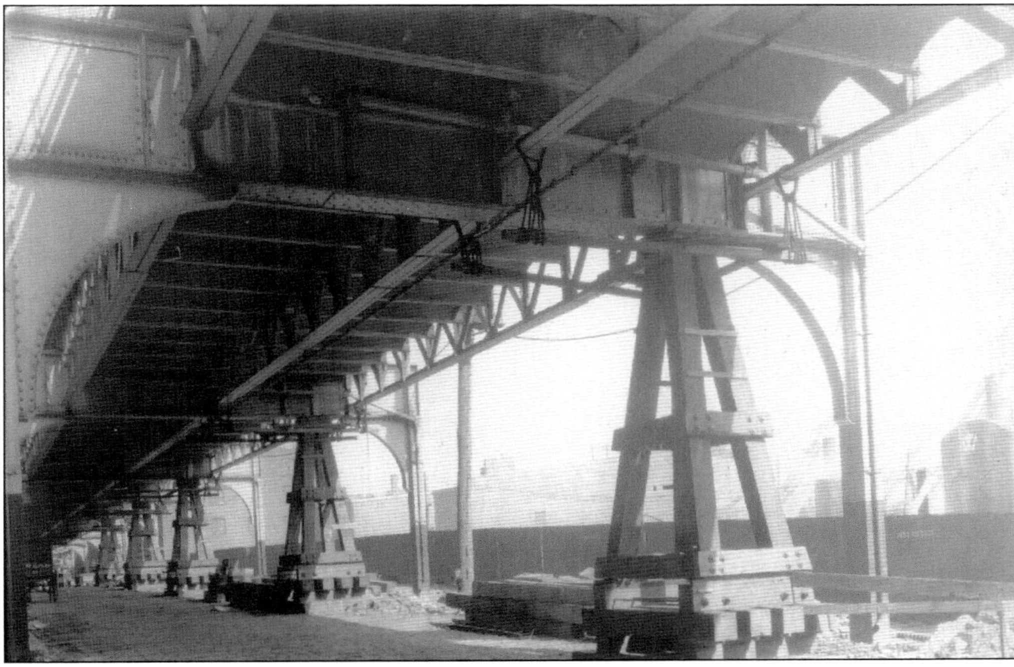

In close proximity to the Philadelphia anchorage lays the Philadelphia Rapid Transit Company's Frankford-Elevated Railroad, forerunner to the Southeastern Pennsylvania Transportation Authority (SEPTA) Market-Frankford Elevated. Since the elevated structure is adjacent to the anchorage site, temporary supports were installed to ensure its stability during excavation. The anchorage site sits just beyond the timber-plank wall beside the sidewalk.

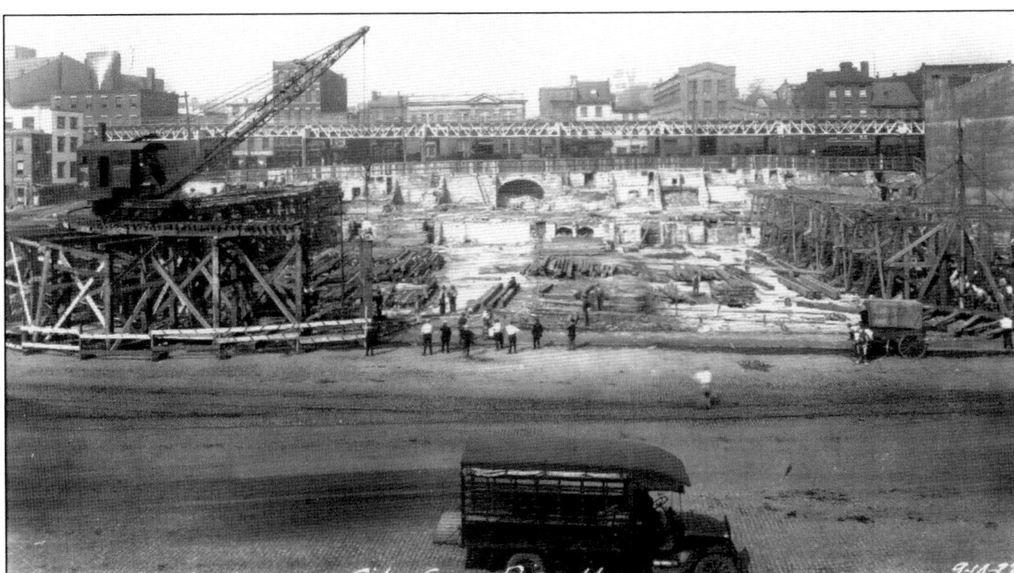

With demolition at the Philadelphia anchorage site complete, work crews erect a temporary timber trestle around the work site. The trestle structure was used by cranes to access all areas within the anchorage site during excavation.

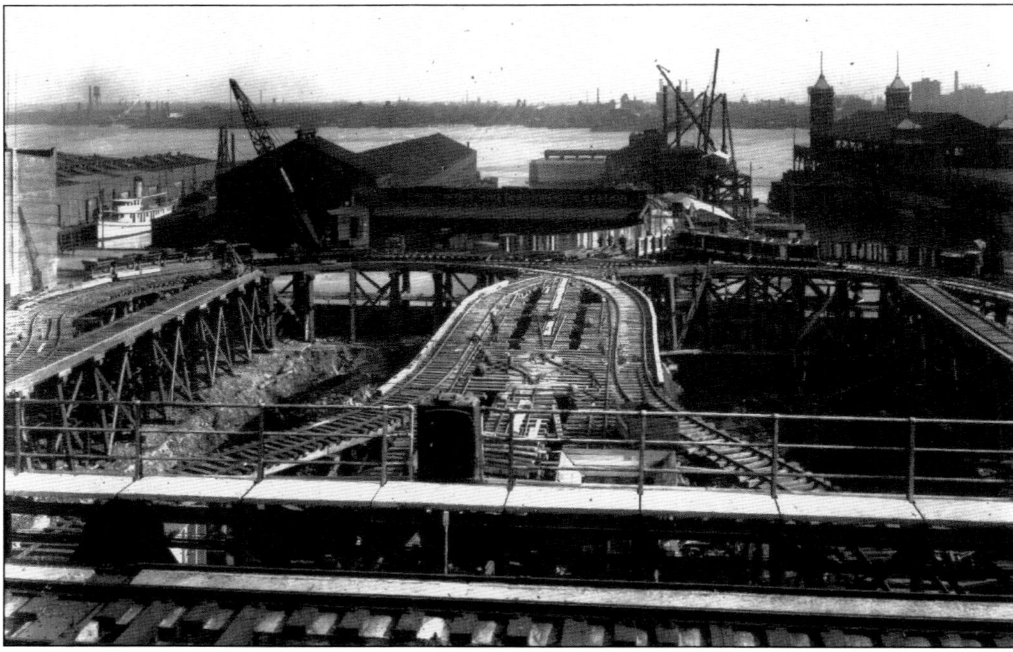

This view, taken from the Market-Frankford Elevated structure, shows the completed wood trestle. Work crews are excavating the site for caisson construction. The caisson is a steel form that is filled with concrete to create the footing to support the anchorage. In the background, along the river edge, work on the Philadelphia tower caisson is underway.

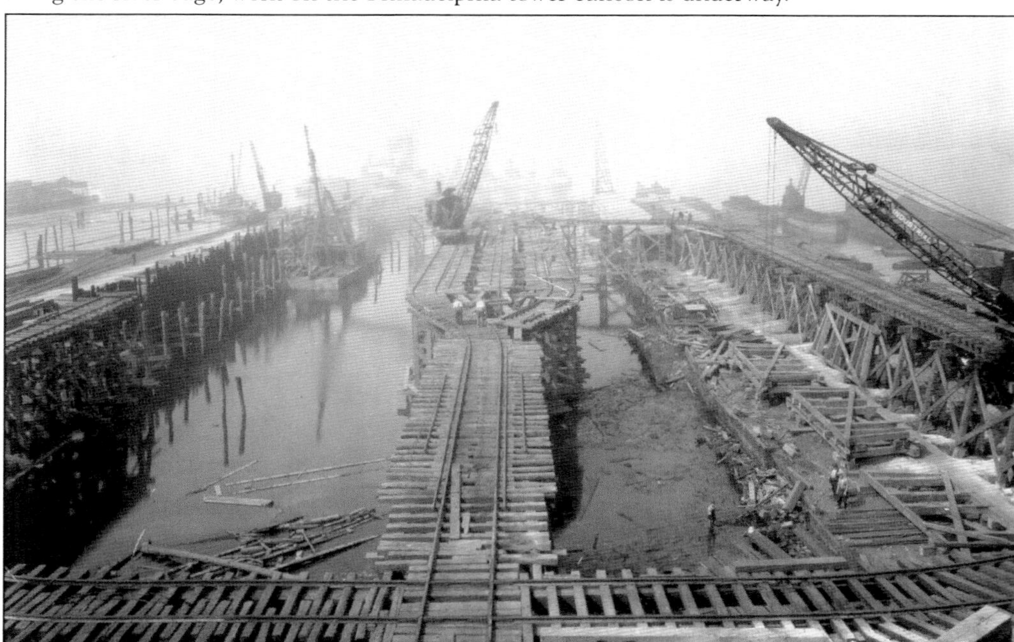

At the Camden anchorage site, a similar trestle structure was erected. Due to its close proximity to the river, a bulkhead was constructed to keep out the water. At both anchorage sites, work crews had to pump out the water prior to excavation.

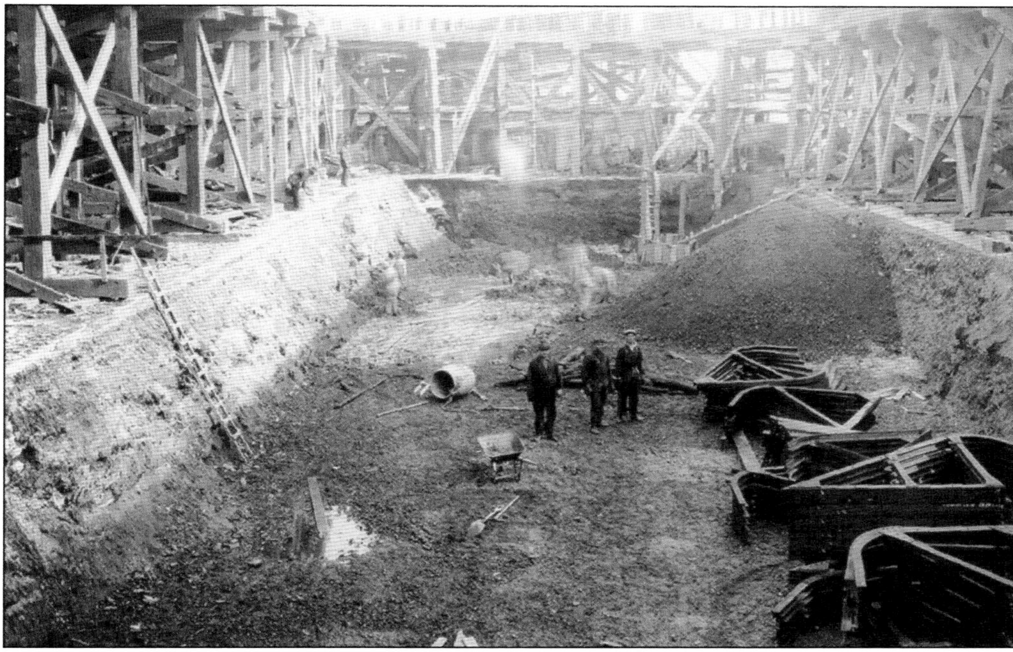

Workmen walk through the excavated work site in Camden before construction of the caissons. The steel seen along the right side of the image will be used to form the cutting edge of two rectangular caissons. The rectangular caissons were composed of multiple hollow cells that allowed cranes to reach into each opening and remove material from the riverbed.

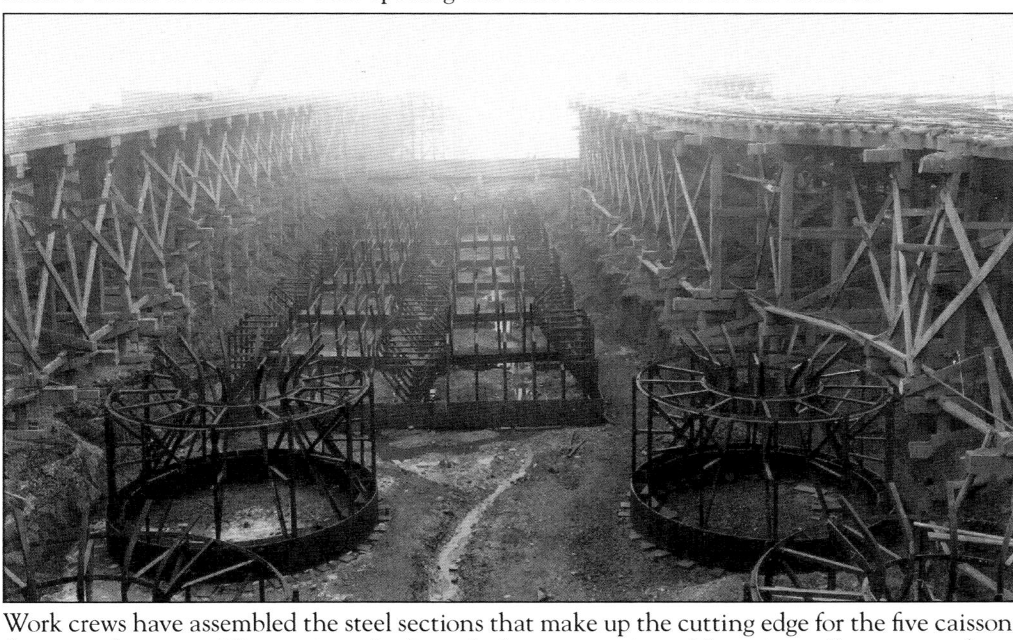

Work crews have assembled the steel sections that make up the cutting edge for the five caissons shown in the image. The cutting edge lines the bottom portion of the caisson. Forms were placed around each caisson to allow concrete to be poured around the steel work. In the background, beyond the rectangular caisson base, a steel bulkhead has been constructed to keep the river out of the construction site.

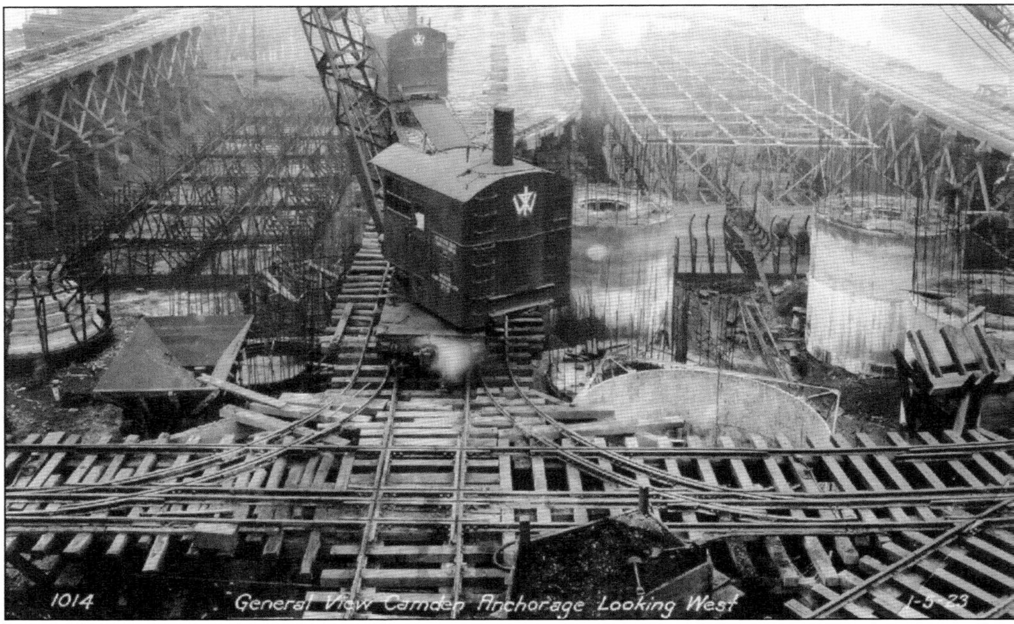

Sinking of the caissons has begun. The three cranes operating at the Camden anchorage removed riverbed material from within each caisson. The circular form seen below and to the right of the crane was placed around the caissons. Concrete was then poured into the form, creating a new layer. Once the concrete had set up, the form was removed and reused. A hopper car full of river material can be seen at bottom (center).

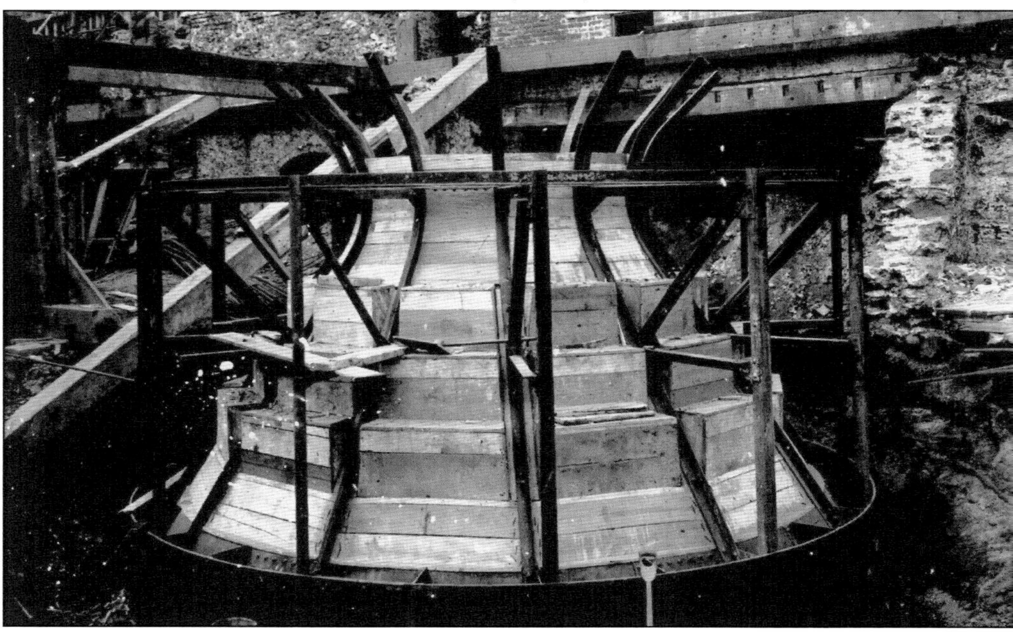

This image shows a close-up of the circular caisson at the Philadelphia anchorage. Workers have installed formwork for the interior face, which was used to create the open chamber inside the caisson after the concrete was poured around the steel skeleton. Inside the chamber, workers dug and removed material as the caisson sunk into the ground.

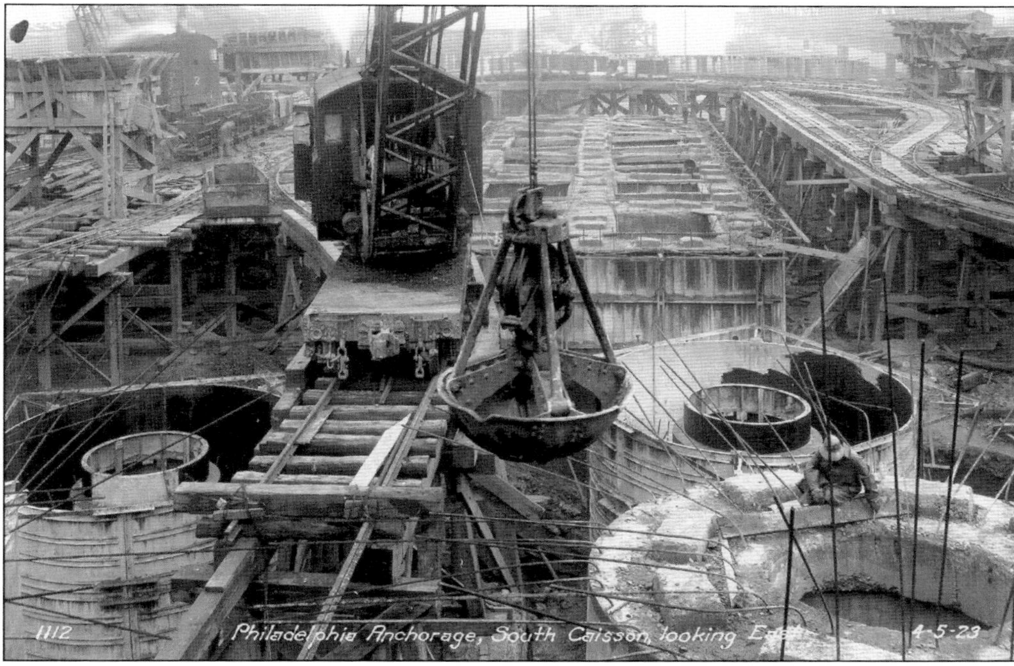

One of the cranes at the Philadelphia anchorage is in the process of removing material from the caisson at the bottom of the image. A worker sits on top of the caisson to direct the movement of the crane and bucket. Reinforcing steel has been bent away from the hole to allow the crane to move the bucket into the working chamber.

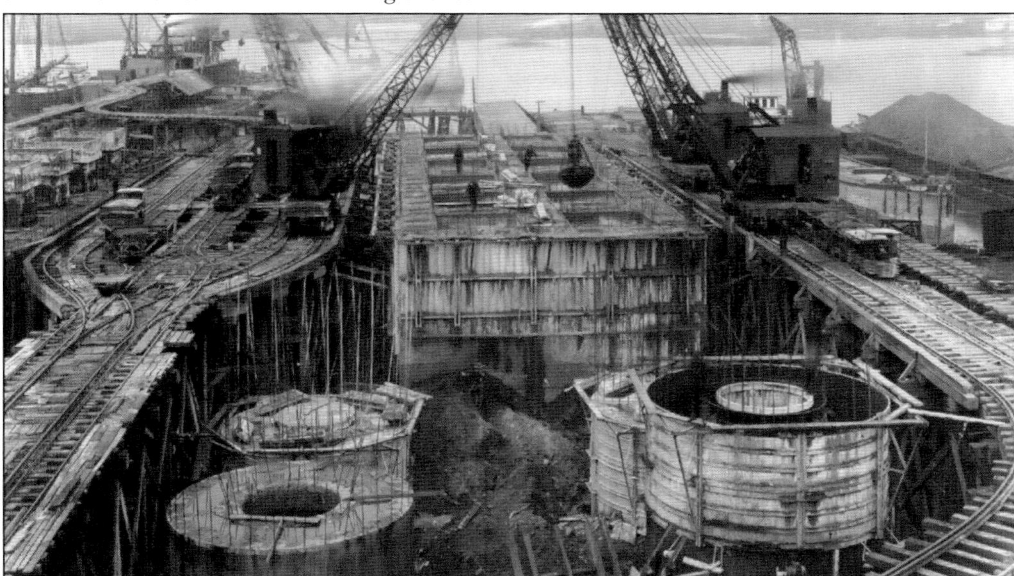

Three cranes at the Camden anchorage are in the process of removing material from the north caisson. A concrete form has been installed on top of the circular caisson in the lower right corner. Once concrete is poured into the form, the added weight, along with the excavated material, helped to sink each caisson. Hopper cars alongside each of the cranes were positioned to take away the excavated material.

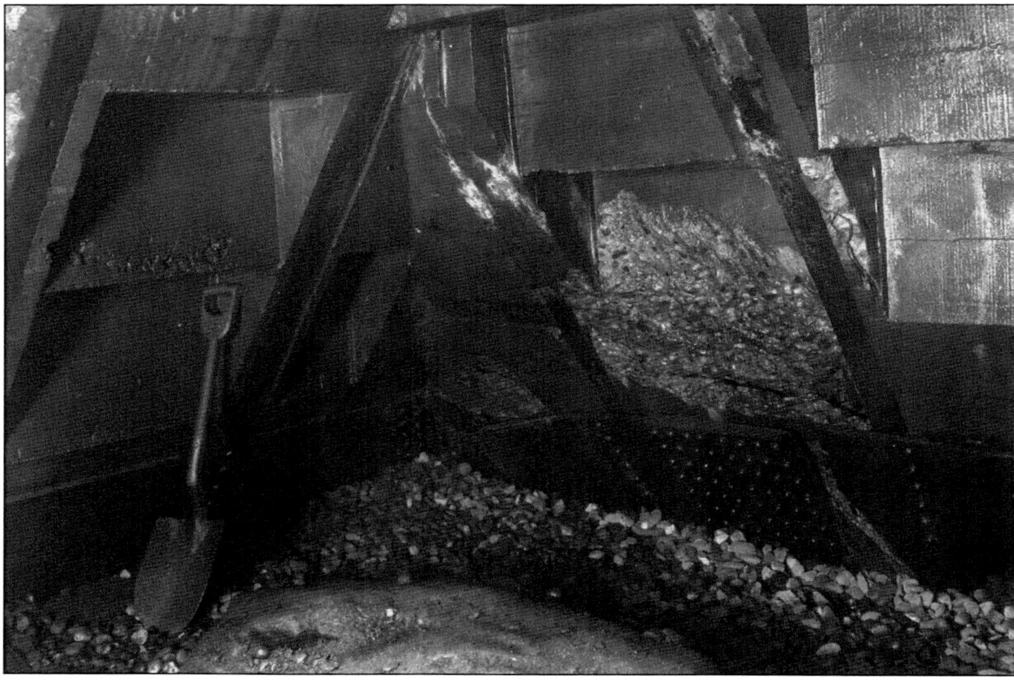

Seen here is the interior of one of the cells in the south rectangular caisson in the Philadelphia anchorage. Although most river debris was removed by the cranes, workers inside this chamber also assisted in removing the material. The steel cutting edge at the base of the caisson is visible.

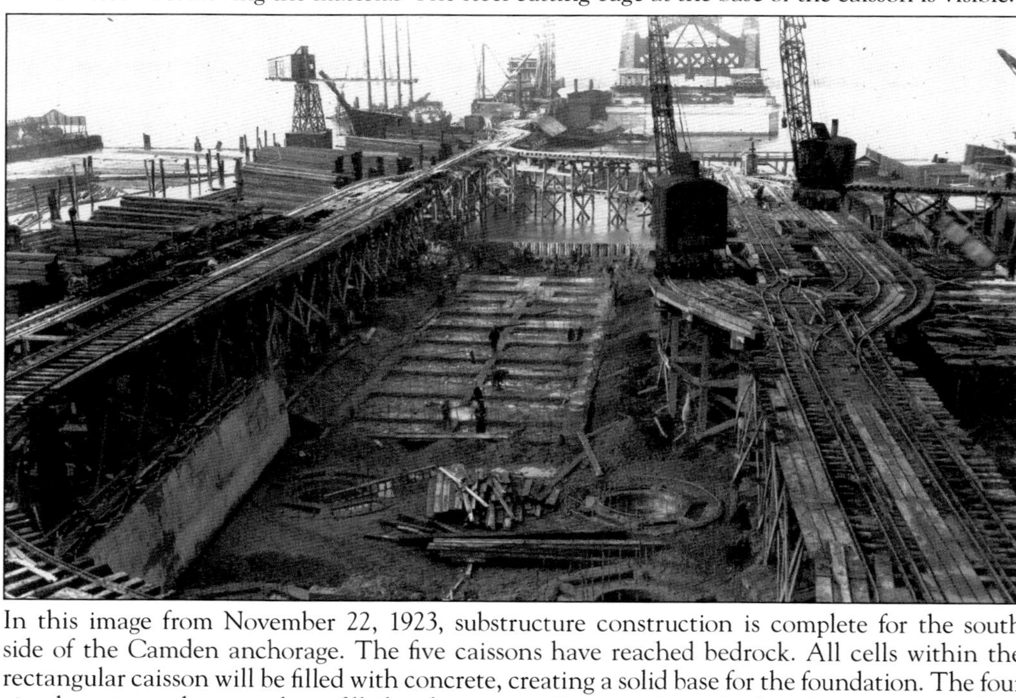

In this image from November 22, 1923, substructure construction is complete for the south side of the Camden anchorage. The five caissons have reached bedrock. All cells within the rectangular caisson will be filled with concrete, creating a solid base for the foundation. The four circular caissons have not been filled with concrete. Once work on the foundation was complete, work on the above-grade portion of the anchorage can begin.

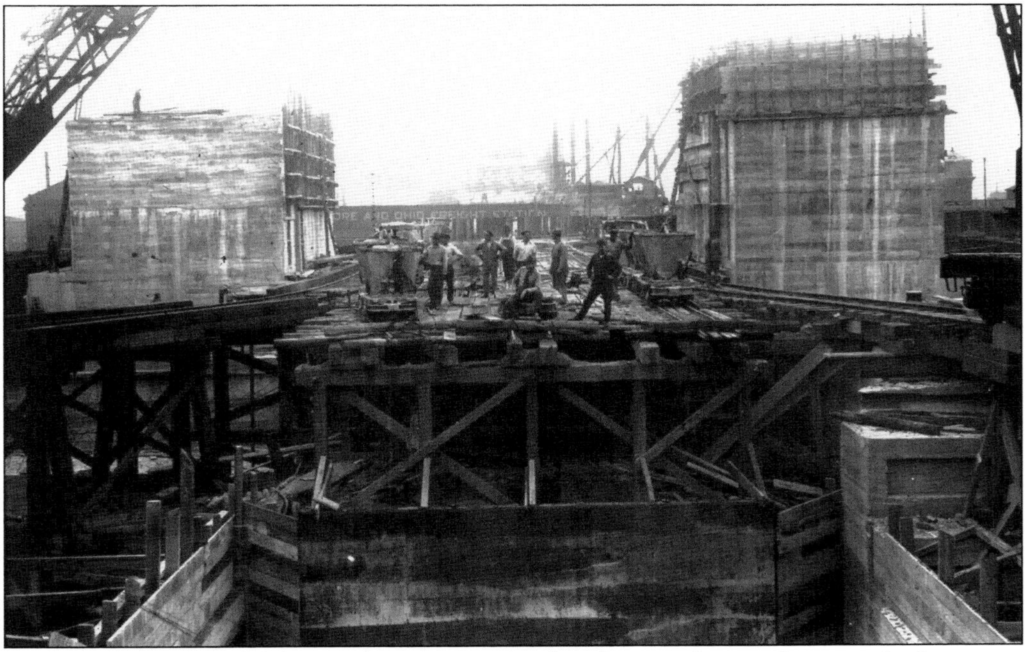

Workers at the Philadelphia anchorage pose for the photographer in this view looking west toward the rising tower. Work on the caissons has been completed, providing a solid base for the anchorage. The two masses of concrete seen rising above both sides of the workers served as the support buttresses for the anchorage's cable bent.

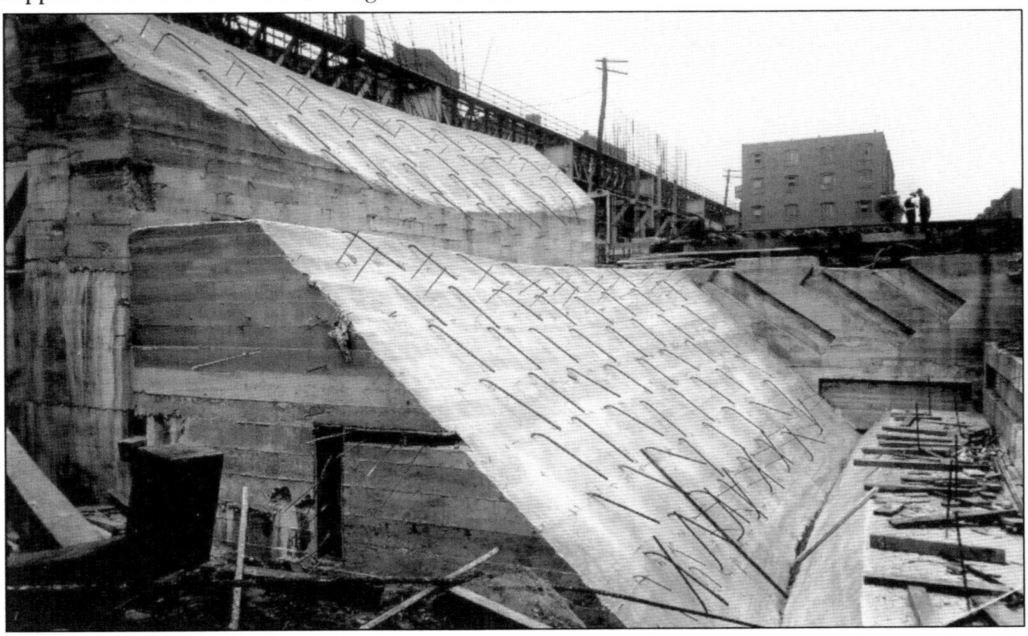

The purpose of the anchorage is to resist the pull of the bridge's main cables. After each cable enters the anchorage, groups of wires (strands) are looped around steel bars (eyebars) embedded within the concrete structure of the anchorage. This image shows the concrete face prior to the eyebar installation. Hundreds of eyebars were used to secure the cables.

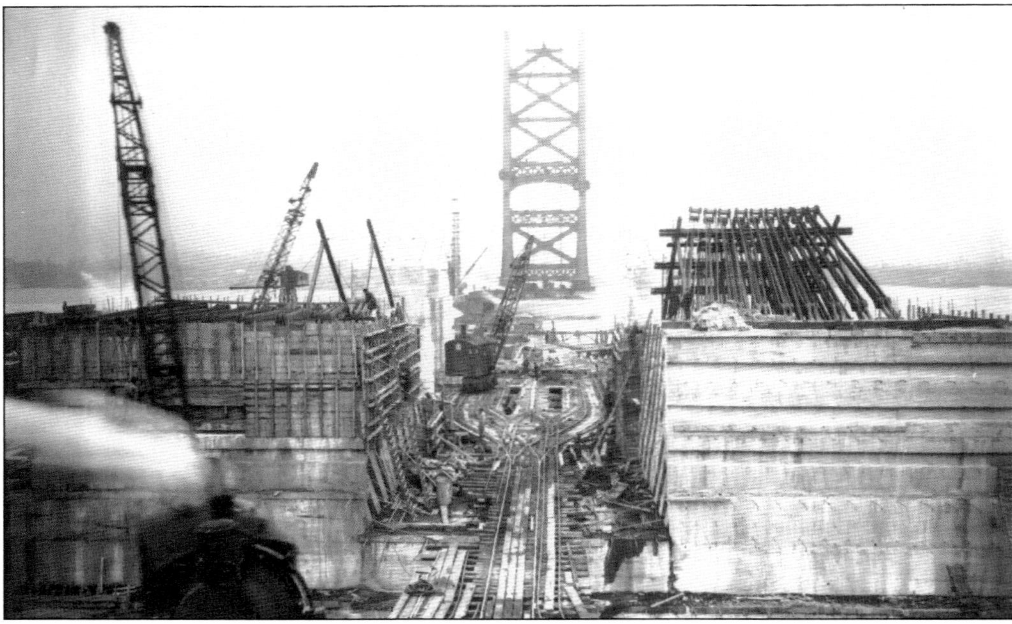

A few months after the previous image was taken, the Camden anchorage shows the eyebars in position for the north cable (right). Workers are moving the eyebars into position for the south cable (left). Once all the eyebars are moved into place, concrete will be poured around them to secure each in place and provide the strength needed to resist the pull of the cable. (Courtesy of Modjeski and Masters, Inc., Mechanicsburg, PA.)

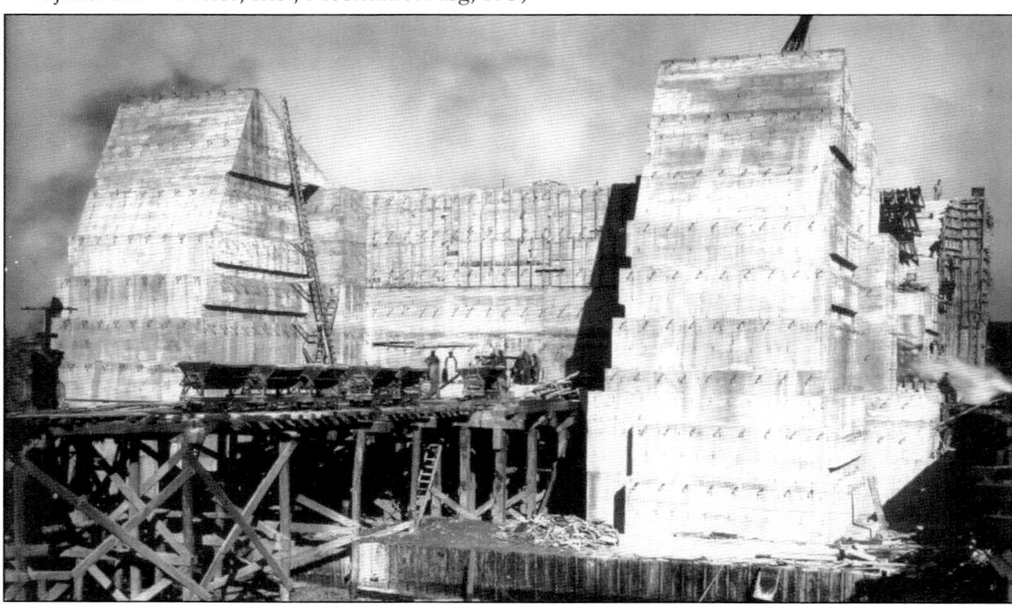

Workers are nearly finished with the first phase of anchorage construction. Eyebars for both cables are in position and embedded within the anchorage concrete. The only remaining item to be installed is the anchorage's steel "bent." The bent will be installed on the two massive concrete buttresses alongside the wooden trestle. Once the bent is moved into position, cable spinning can begin.

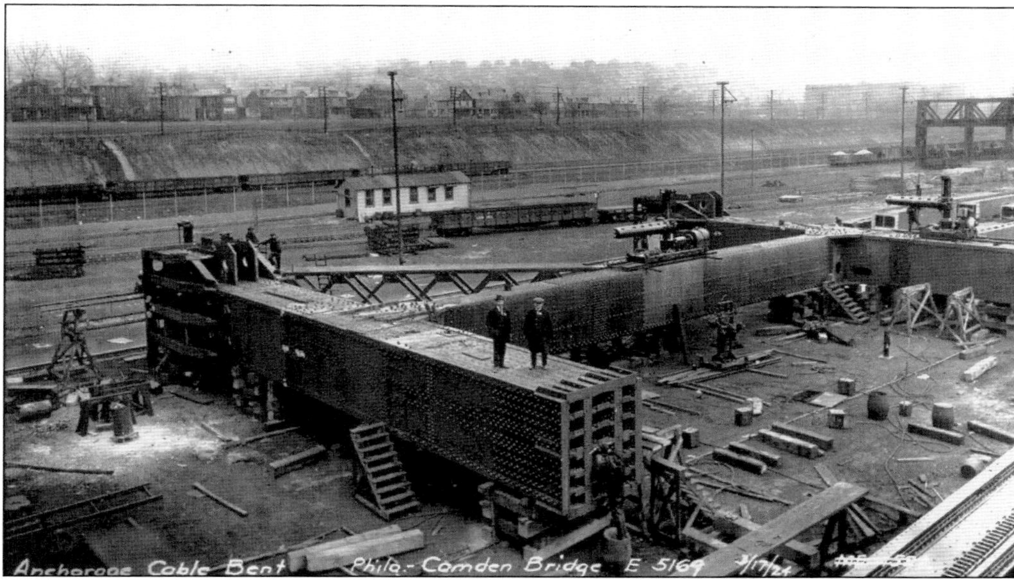

At both anchorages, there is an inclined steel frame called the anchorage bent. As described by the DRBJC assistant engineer at the time of construction, Charles Carswell, "The anchorage bent performs a function [similar to a] backyard clothes prop. The cables rest upon the bent and are deflected downward to their attachment to the anchorage steel [eyebars]." The anchorage bents were constructed by American Bridge Company's Ambridge Shop. As shown in photograph above, the bents were preassembled at the fabrication shop to ensure proper fit. The bent was disassembled and shipped to the construction site for final assembly and installation. In the June 1924 photograph (below), the 900-ton anchorage bents have been assembled on the Camden anchorage. Once the bents are firmly secured and braced, cable stringing can commence.

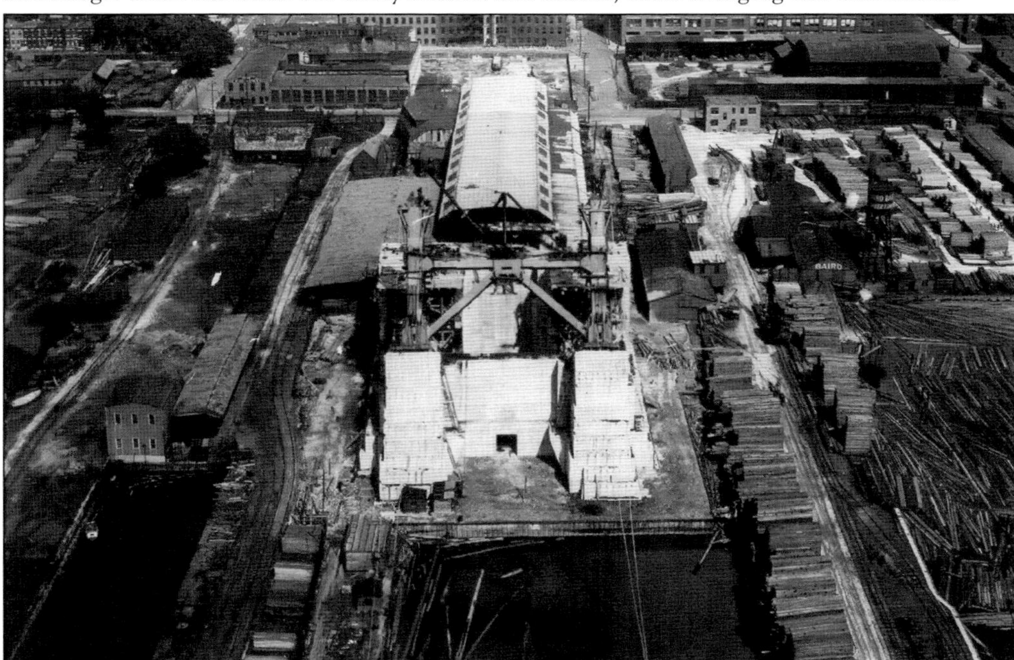

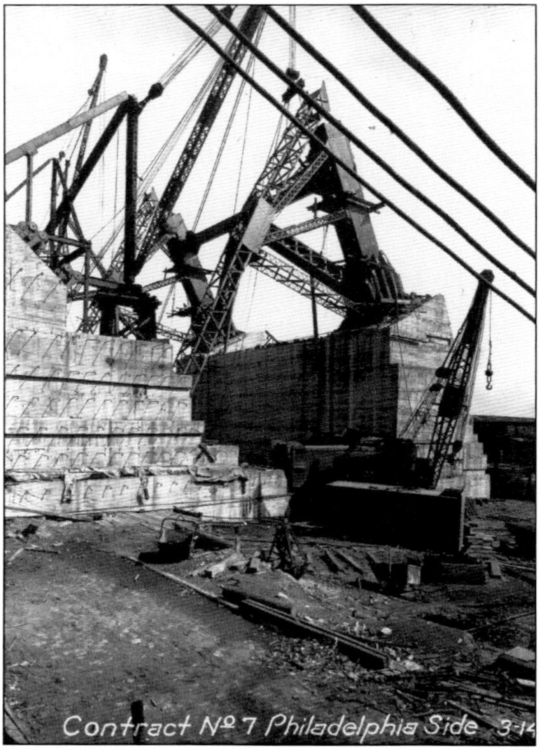

This elevation view of the Philadelphia anchorage shows the anchorage bent in position. Temporary steel bracing has been moved in to secure the bent and keep it from falling back into the anchorage. After the cable stringing is finished, the braces will be removed. Enough force is generated by the cables to keep the anchorage bent in position.

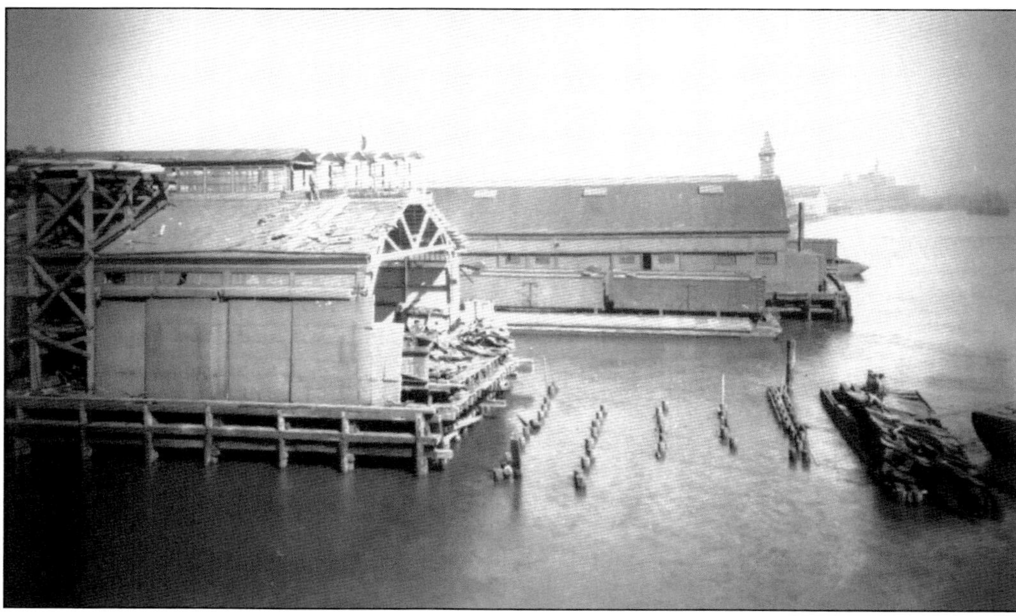

One of the Department of War's requirements for the Delaware River Bridge was that the towers had to be located within the pierhead lines. Because of that stipulation and further recommendations regarding bridge location, the Philadelphia tower fell within the footprint of Pier 11, seen here. As such, the pier had to be partially demolished in 1922 to make way for the new tower. (Courtesy of Modjeski and Masters, Inc., Mechanicsburg, PA.)

Workers at New York Shipbuilding Corporation in Camden are seen here constructing one of the caissons used to create a tower foundation. Each caisson has a footprint of 79 feet by 143 feet; the overall height of the two caissons differs between 65 feet (Philadelphia) and 89 feet (Camden) due to the depth of bedrock at each site. After construction, each caisson was slid into the river and floated into position.

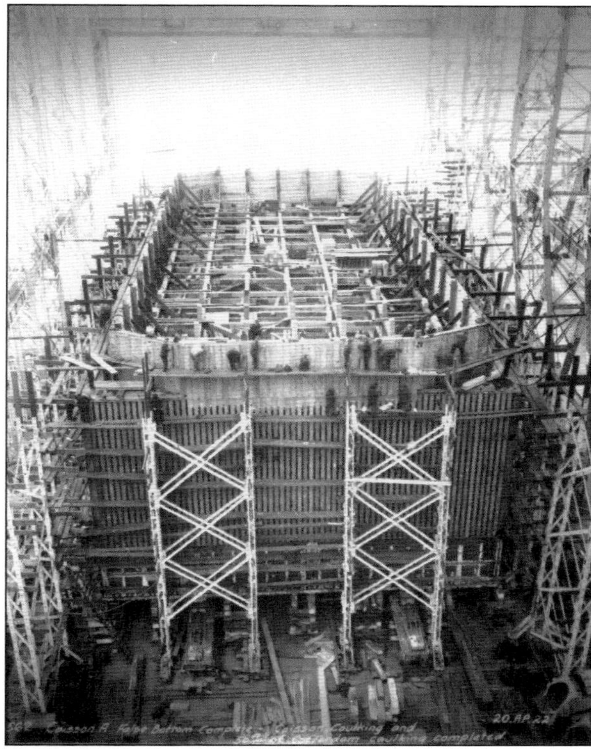

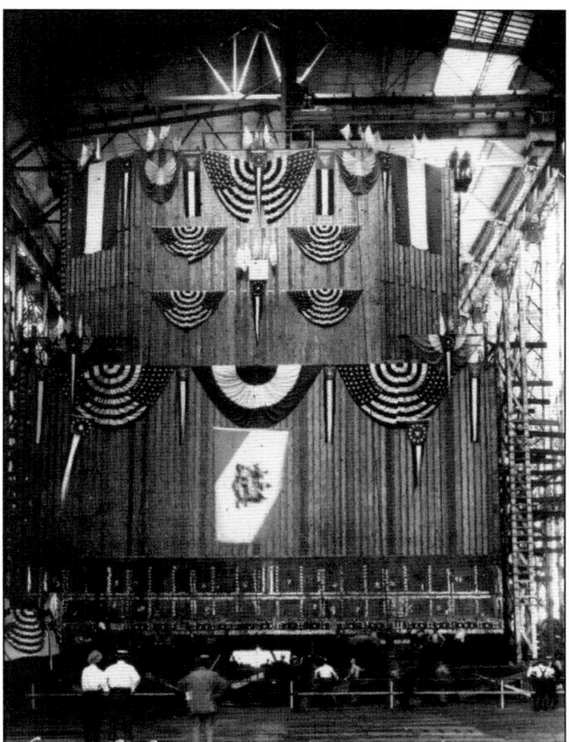

Once the Camden caisson was completed, it was decorated with flags and bunting. At the bottom of the caisson, the outside face of the steel "cutting edge" can be seen. This edge was used to cut into the riverbed during the sinking process. The overall scale of the caisson can be seen by comparing it with the people milling around the steel base.

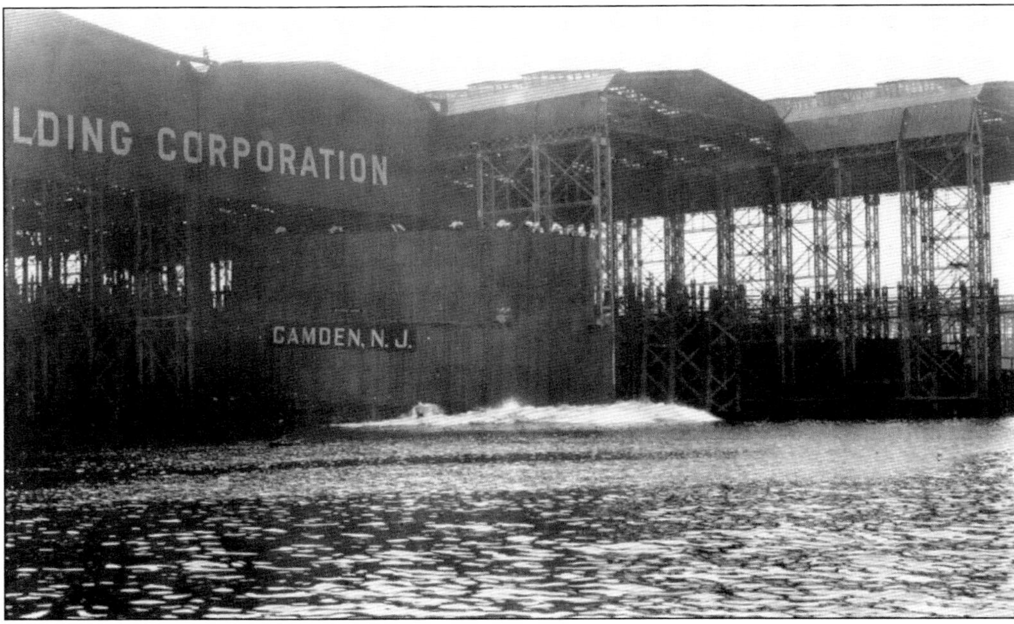

The Philadelphia caisson was launched on April 24, 1922; this image shows the Camden caisson being launched into the Delaware River on July 18, 1922. Eleven days after launching, the caisson was floated upriver into position at the tower site.

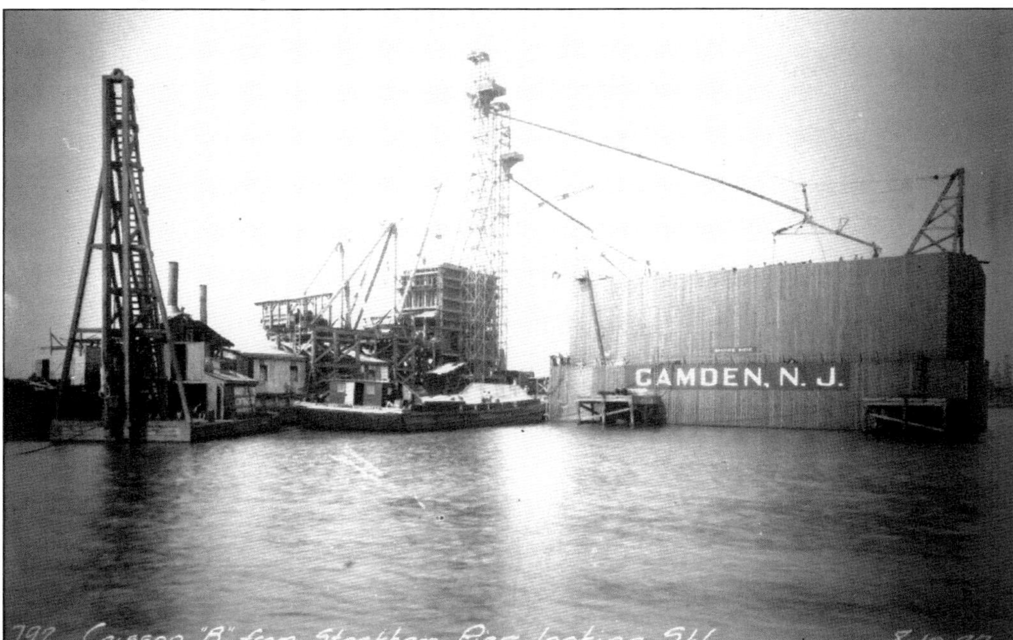

Once in position, the Camden caisson was sunk to the river bottom. Compressed air was supplied to the working chamber at the base of the caisson, allowing men called "sand hogs" to work there safely. Working within the chamber, sand hogs removed riverbed material while concrete was added to the caisson above, allowing the caisson to sink into the riverbed at an average rate of 15 inches per day. After six weeks, the caisson reached bedrock.

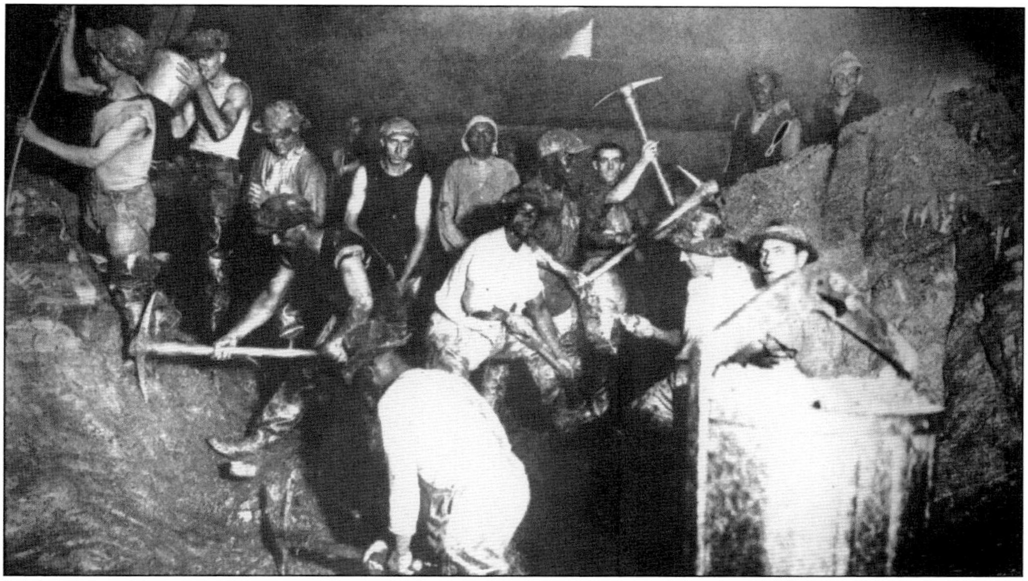

River-bottom material beneath each tower foundation was removed by hand. The sand hogs toiled in a pressurized air chamber at the base of each caisson. As river debris was removed and concrete poured into the caisson above the work chamber, the additional weight forced the caisson deeper into the river bed. Once bedrock was reached, the working chamber was closed off and sealed with concrete. According to Charles Carswell, "Inside the working chamber, conditions are very different from above ground. The water is kept down level with the bottom of the caisson walls, or 'cutting edge,' by air pressure, and the men worked in the dry, shoveling material into buckets to be hoisted to surface through air locks." The men often worked without shirts even in winter due to the higher temperatures inside the chamber. Precautions such as having doctors on site were taken to protect workers from "the bends," a possibly fatal condition caused by emerging too quickly from an atmosphere of compressed air pressure.

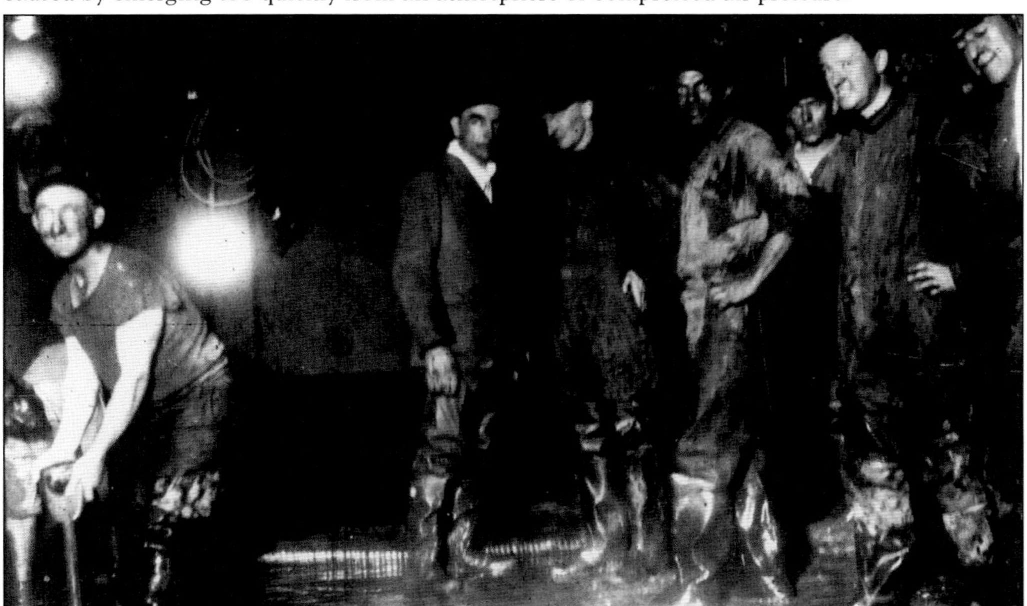

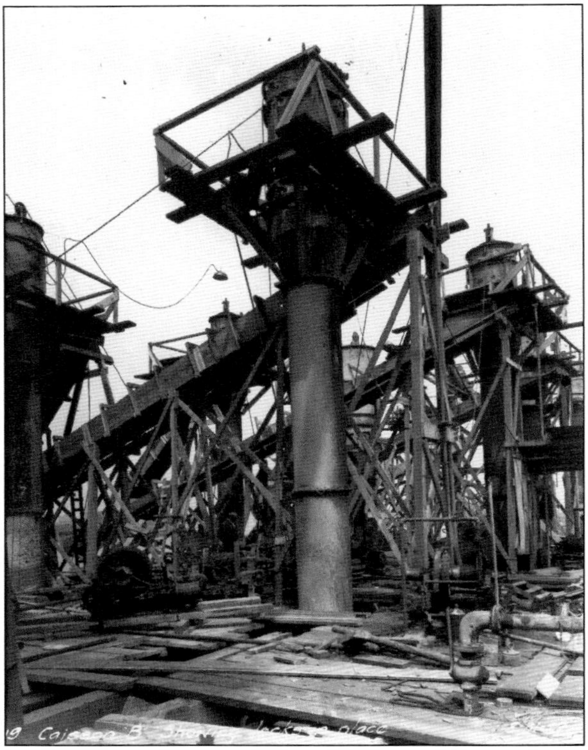

At the end of August 1922, the sand hogs had just begun the process of removing river spoils from the working chamber. Since the sand hogs worked in a pressurized environment, material was hauled out of the chamber through air locks shown in this picture. Several locks were used to remove debris without depressurizing the chamber. Once material was removed from the locks, it was deposited onto the adjacent chute for disposal.

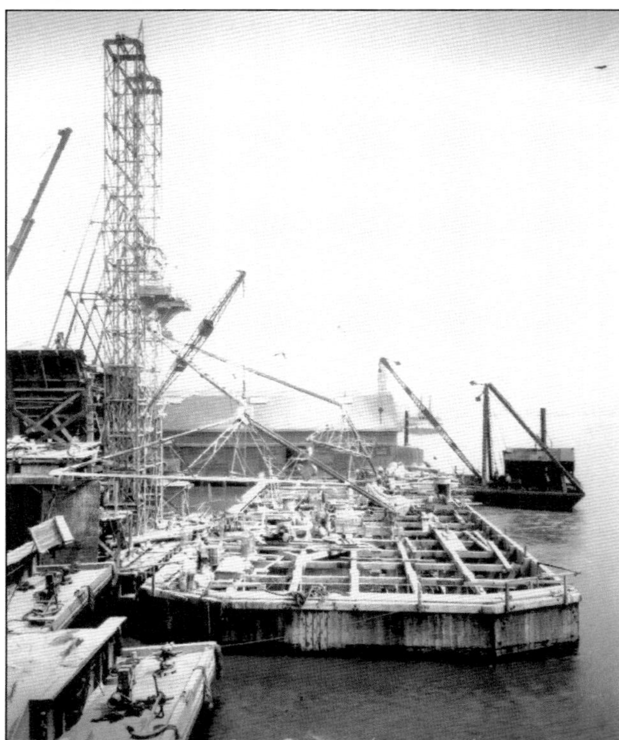

After five weeks of work, sand hogs at the Philadelphia site removed enough material with concrete added above to allow the caisson to sink at an average rate of seven and a half inches per day. Five days before this photograph was taken, the caisson reached bedrock. Concrete was poured into the working chamber to seal the structure to bedrock. (Courtesy of Modjeski and Masters, Inc., Mechanicsburg, PA.)

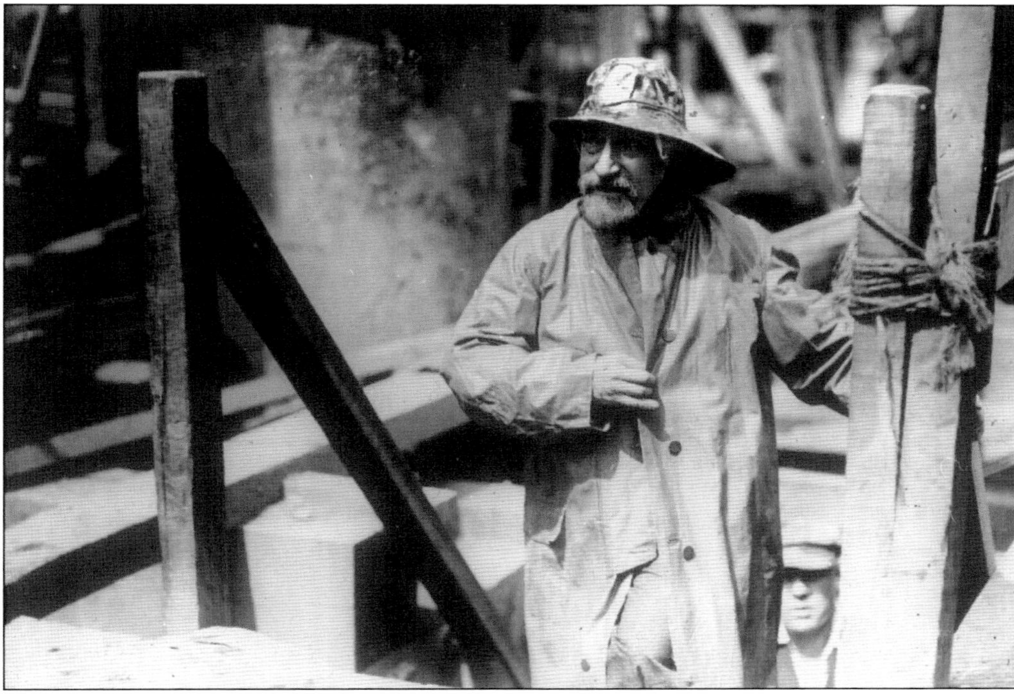

In June 1922, chief engineer Ralph Modjeski emerges from the Philadelphia caisson after an inspection of the working chamber at the base of the caisson.

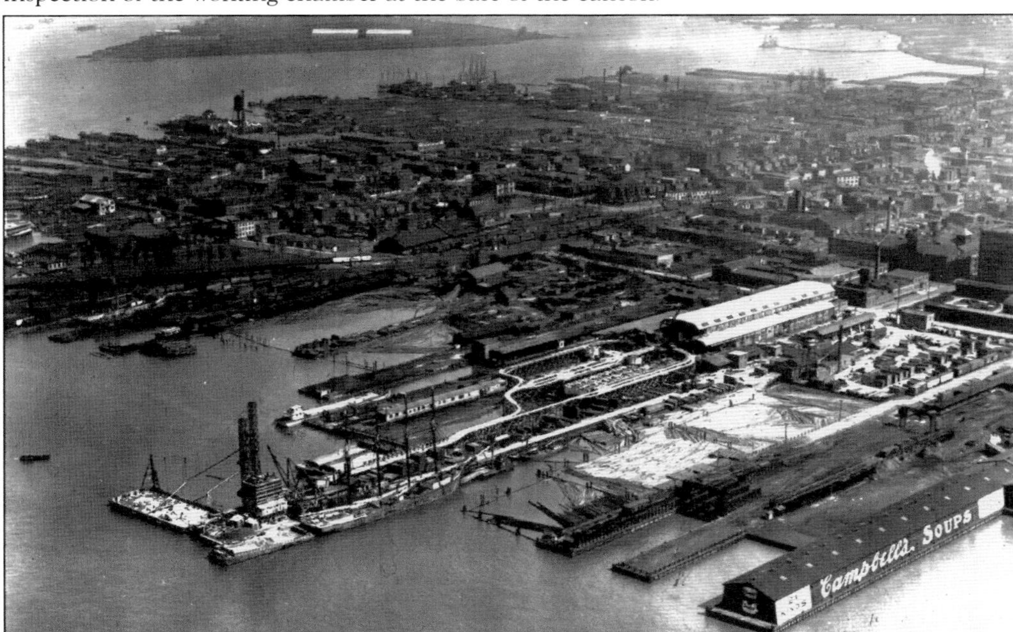

This aerial photograph of the Camden site shows work progressing on the tower and anchorage caissons. The long white building to the right of the Camden anchorage site is the bridge's wire plant. According to the advertisement on their pier, Campbell's Soup featured up to 21 kinds of soup at this time.

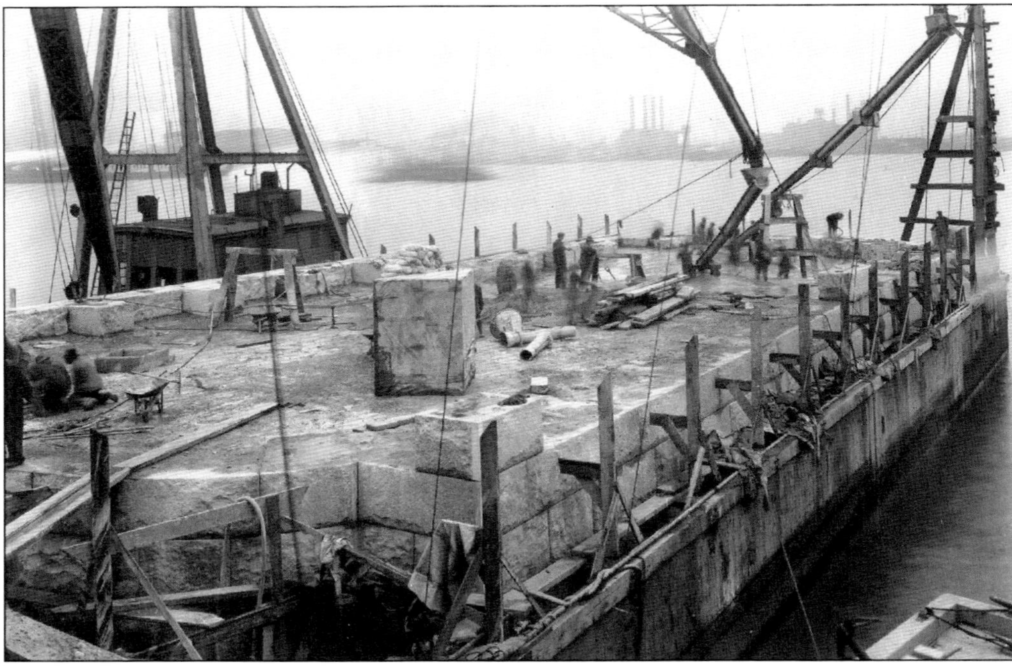

After the caisson reached bedrock and was sealed off, workers began to complete the tower foundation or "pier." Granite blocks were installed around the exterior face of the tower pier. The granite blocks were quarried at Stone Mountain, Georgia. Each block was cut to size at the quarry and shipped to the pier site for installation.

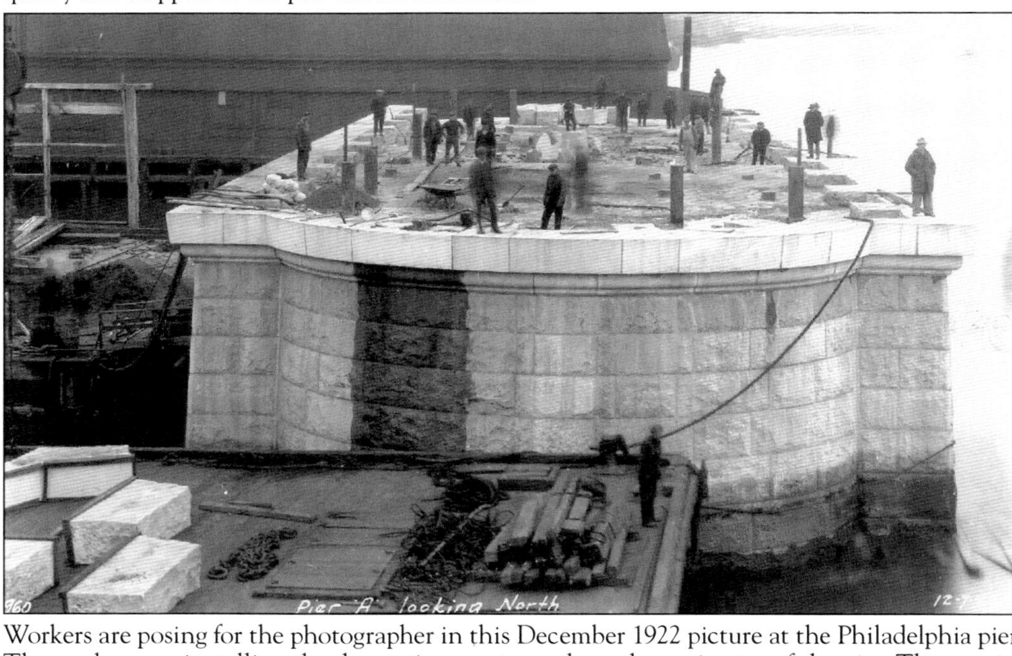

Workers are posing for the photographer in this December 1922 picture at the Philadelphia pier. The workers are installing the decorative capstone along the perimeter of the pier. The granite was transported to the construction site by barge, as seen at the bottom of the image, and lifted onto the pier by crane.

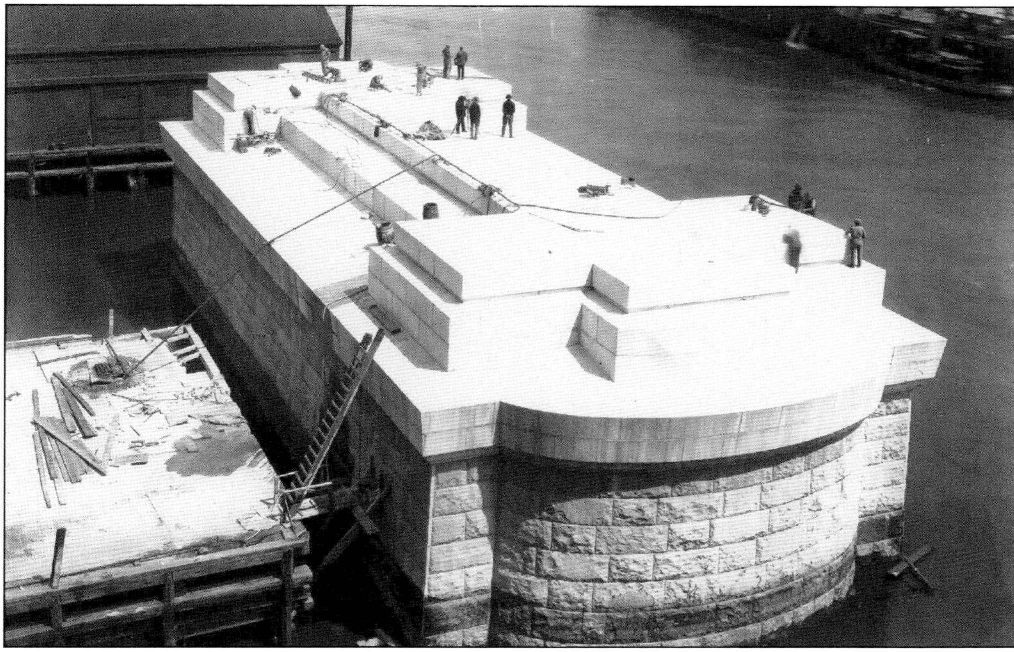

The setting of granite blocks for the Philadelphia tower pier was completed in March 1923. On top of the pier are three courses of granite blocks that make up the bearing seats for the tower legs. After the top course of granite was installed, the entire surface was planed to a slope not exceeding 1/32 of an inch. Grooves were also cut into the granite to ensure a proper seating for the tower steel.

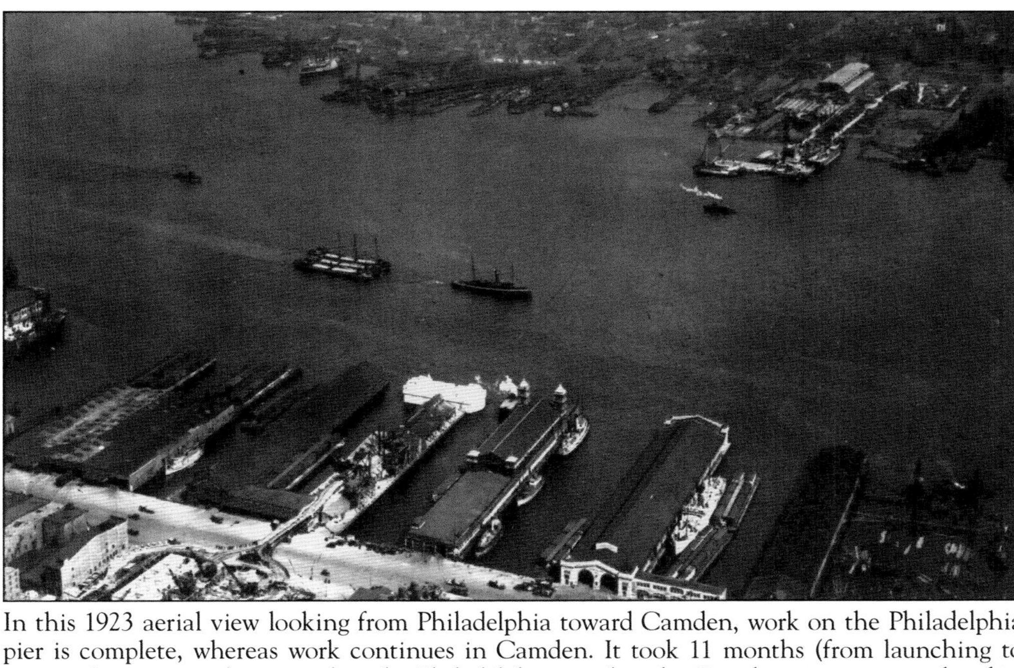

In this 1923 aerial view looking from Philadelphia toward Camden, work on the Philadelphia pier is complete, whereas work continues in Camden. It took 11 months (from launching to setting the capstones) to complete the Philadelphia pier, but the Camden pier was completed in 10 months, even though the pier had to be sunk to a depth greater than the Philadelphia pier.

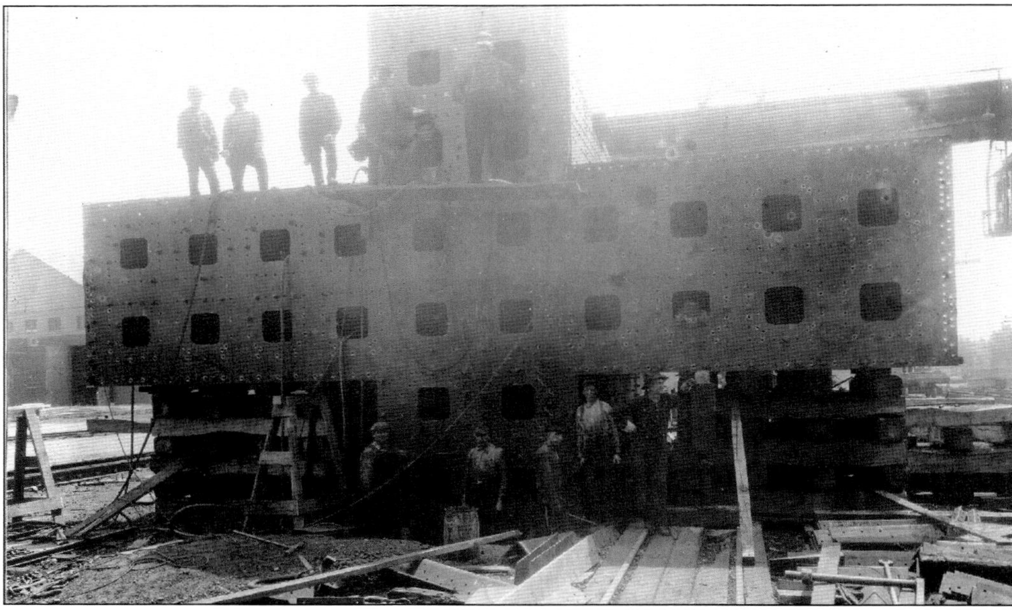

At Bethlehem Steel's Steelton plant, workers pose alongside the base plate of the tower leg. The cellular construction of the tower allows it to be flexible when loads are applied to the bridge. The open cells also provide a means for accessing and maintaining the various portions of the tower and reducing its overall weight.

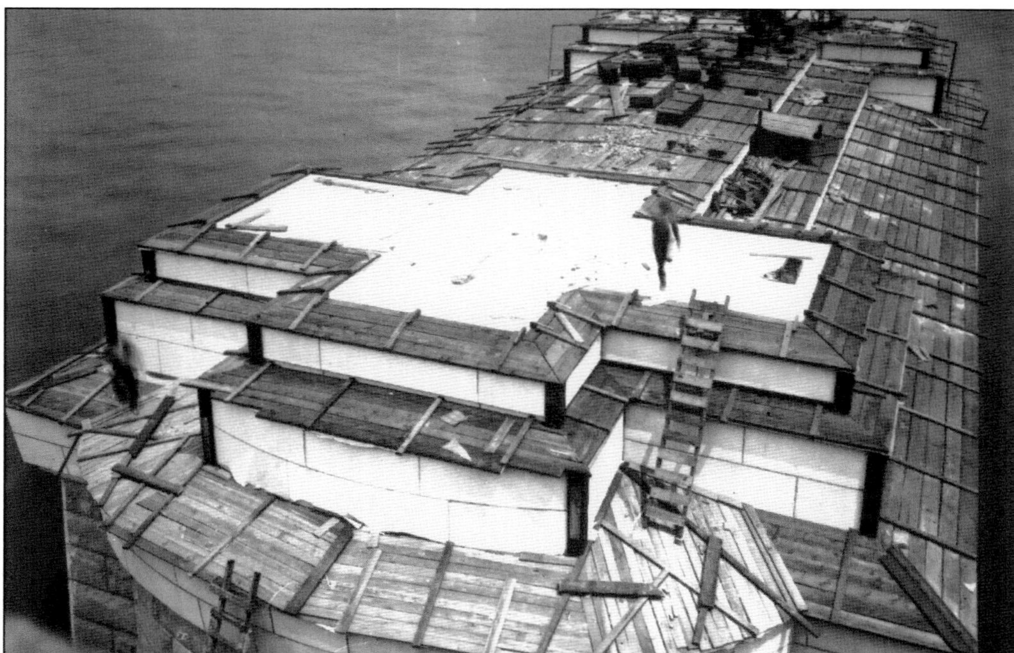

Prior to grinding the granite surface of the Camden pier, workers have installed wood planking along the top of the pier. The exposed surface will be planed to a level surface. Grooves will be etched into the surface to assist with the base plate installation. (Courtesy of Modjeski and Masters, Inc., Mechanicsburg, PA.)

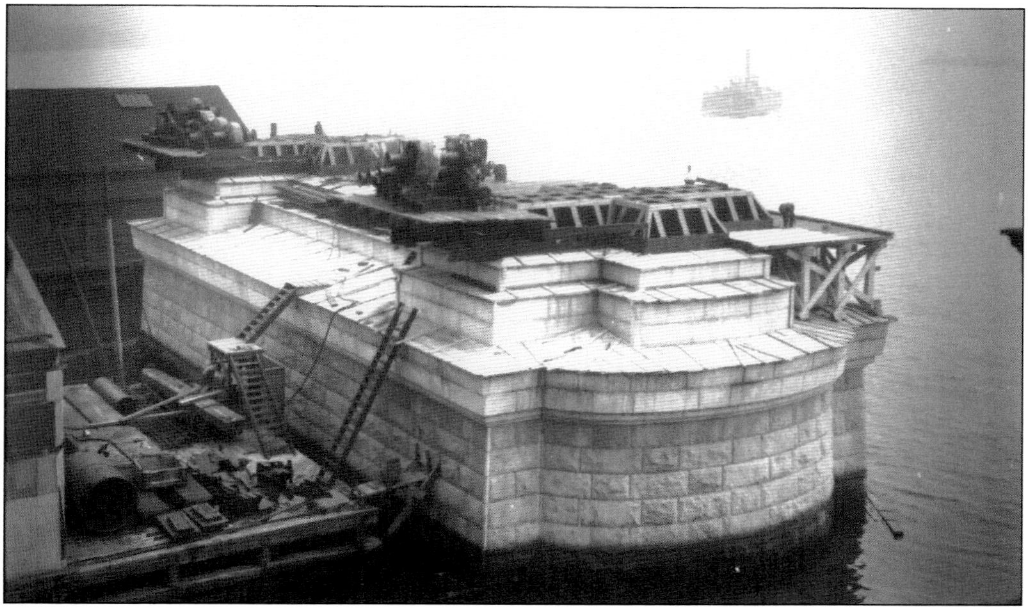

In August 1922, workers began installing the tower's leg base plates on the Philadelphia pier. Although the granite surface and base plates were planed to create a level surface, the workers poured a red lead paste beneath the plates before the base was set. After the base was set, it was lifted out to inspect the paste. Workers found that the paste filled in all the surface irregularities, ensuring that water voids had been forced out from between the plate and the granite. (Courtesy of Modjeski and Masters, Inc., Mechanicsburg, PA.)

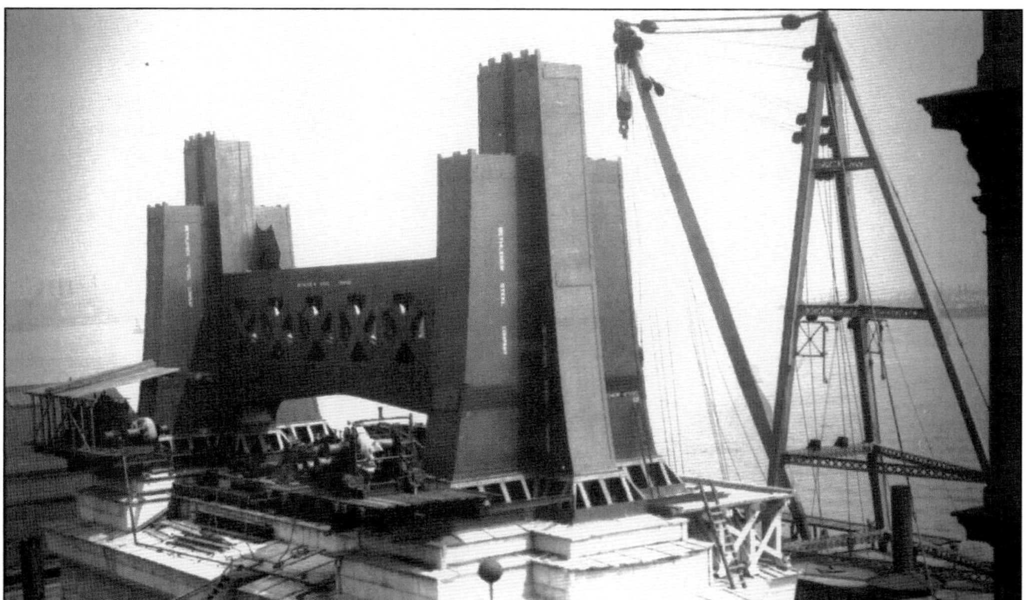

Installation of the first tower leg sections begins at the Philadelphia pier. The first steel sections were set into position with a floating derrick crane seen to the right of the tower pier. Each tower section was shipped to the tower site by barge, which was moored to the pier as the sections were lifted into position by crane. Some of the tower sections weighed up to 50 tons.

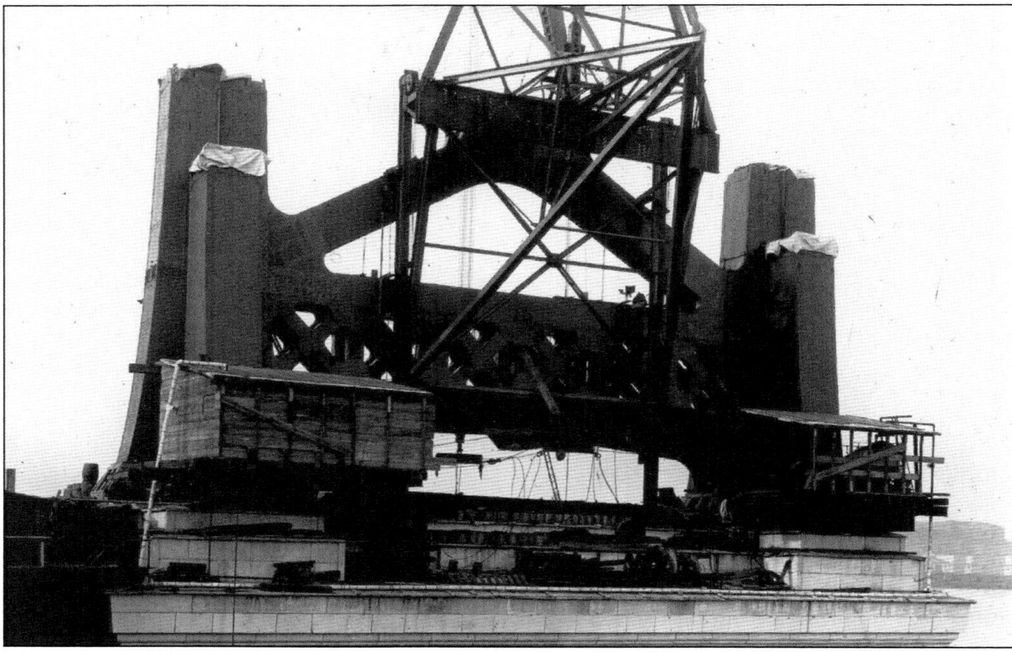

This image of the Camden tower shows the installation of traverse bracing beneath the roadway level. Traverse bracing is used to connect the tower legs, creating a single, rigid unit. A "creeper traveler" was used to move the steel sections from the barge to the tower. The creeper traveler was a crane secured to the tower; as the tower grew higher, the crane lifted itself along guide beams attached to the tower's upper sections.

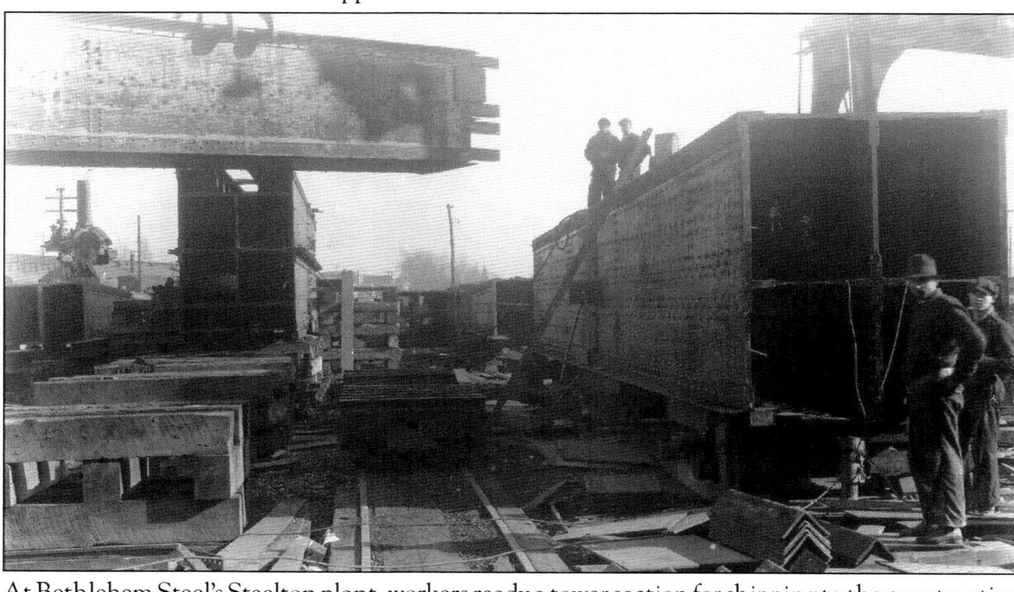

At Bethlehem Steel's Steelton plant, workers ready a tower section for shipping to the construction site. Tower sections were preassembled at the site to ensure a proper fit. Each bearing surface was planed to a maximum deviation of 1/100 of an inch from true vertical. The high degree of precision was necessary for the tower sections to fit properly; any deviation in alignment at the tower base would have an adverse impact at its top.

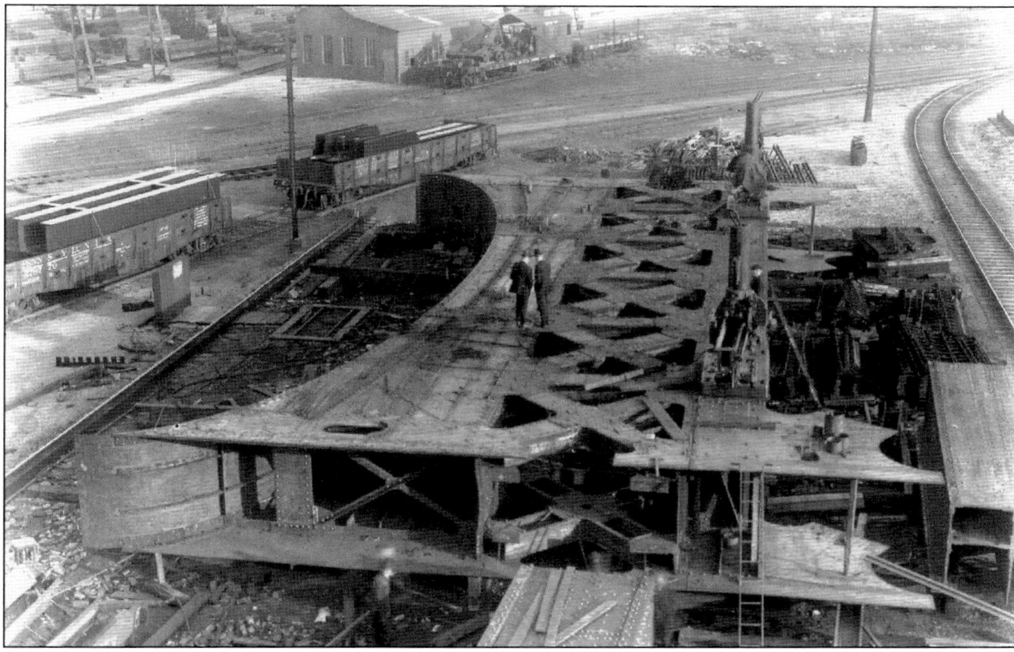

The Steelton shops are in the midst of completing the fabrication of the roadway portal strut for the Camden tower. The scale of this section of the tower is evident when compared with the workers walking across the portal strut's face. As can be seen in the background, smaller tower sections have been loaded onto flatbed railroad cars ready for shipping to the construction site.

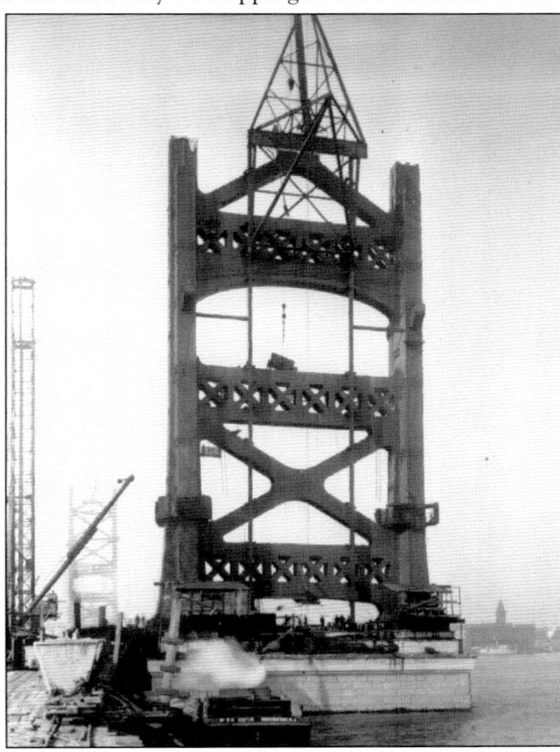

Work on the Camden tower has progressed to the lower traverse bracing above the roadway level. The roadway portal strut seen in the prior image has been installed. In the background, work on the Philadelphia tower is nearing completion as crews are installing the top traverse bracing. (Courtesy of Modjeski and Masters, Inc., Mechanicsburg, PA.)

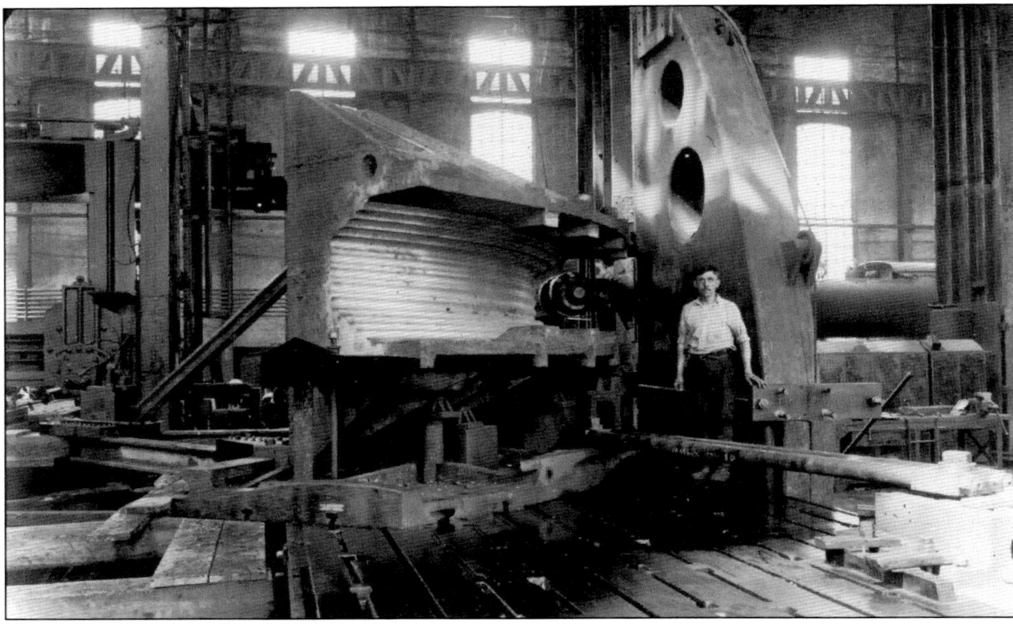

At Bethlehem Steel's Bethlehem plant, work is progressing on the machining of the tower saddle casting. This casting was installed on top of each tower column. The saddle allows the cable to be "draped" over the top of the tower, as indicated by Charles Carswell's description of it as being "like a prop for a clothes line."

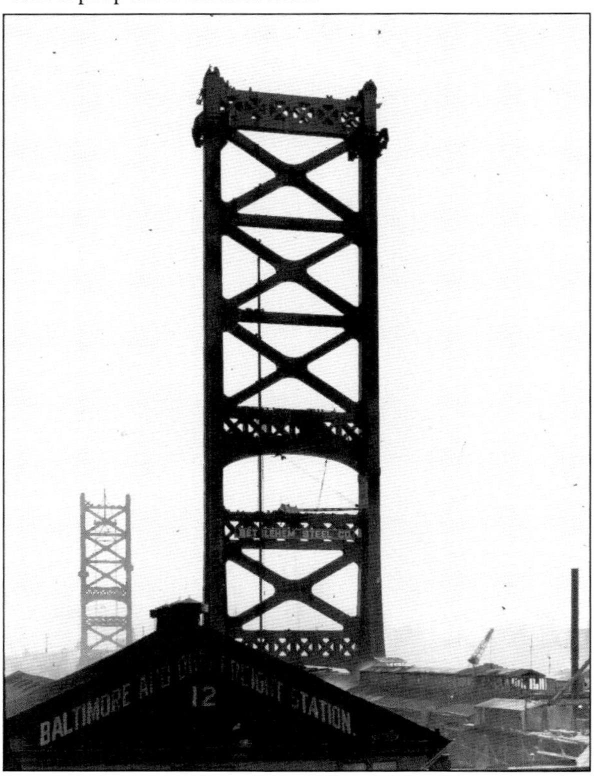

In this March 1924 image, the Philadelphia tower is nearing completion. The Camden tower in the background is two months away from being finished. Although the tower sections were installed with great precision, the Camden tower did develop a twist. During construction, the top of the Camden tower had twisted two and three-quarter inches. Engineers determined that the slight rotation was acceptable and could be fixed during cable stringing.

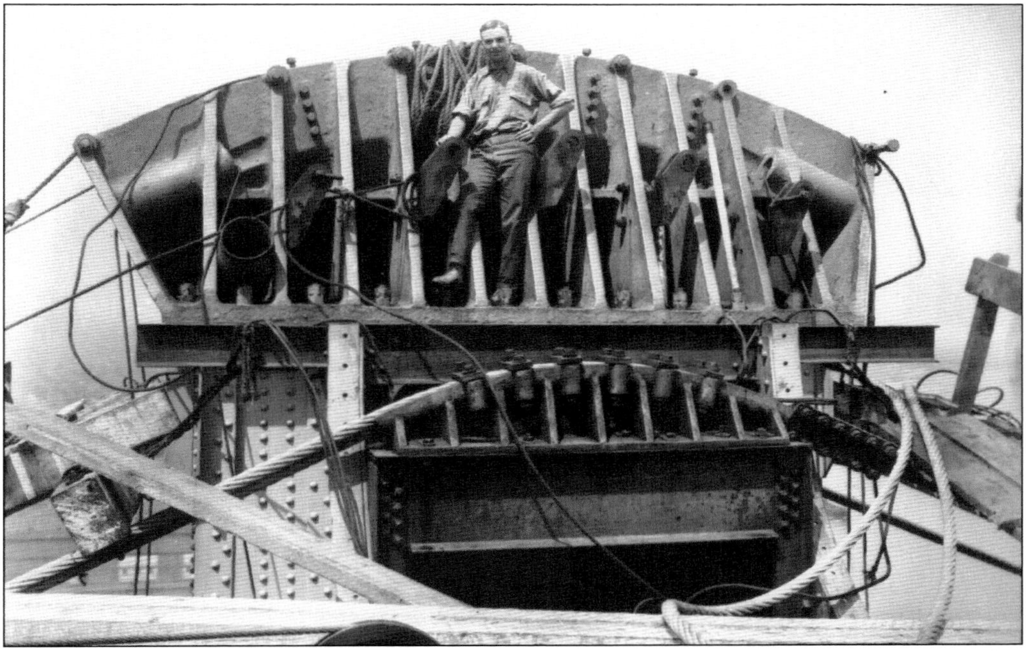

A worker poses alongside the cable saddle installed on top of the Philadelphia tower. After stringing of the main cable, the saddle was enclosed to protect the cable from the elements. (Courtesy of Modjeski and Masters, Inc., Mechanicsburg, PA.)

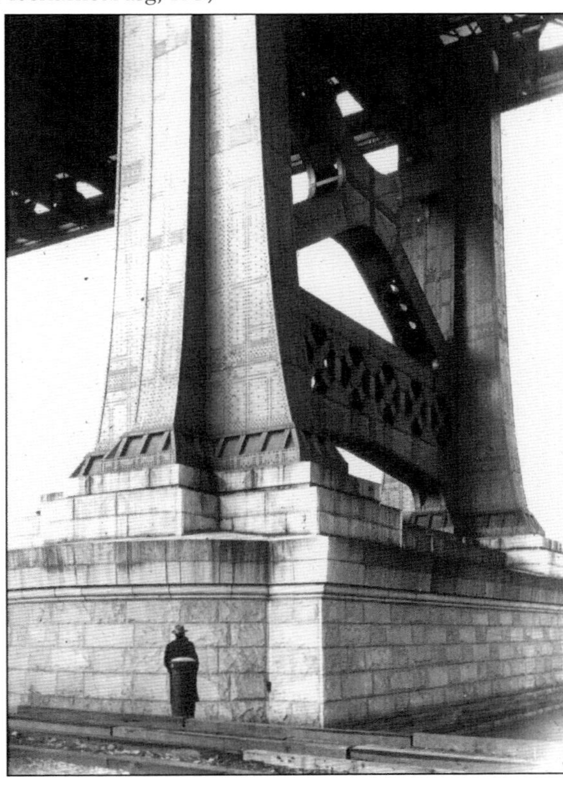

A lone individual, standing on the David Baird Company lumberyard pier, looks over the completed Camden tower. The size of the tower's pier alone is overwhelming. Work on the towers was completed in May 1924, almost two years after the caissons were launched from the New York Shipbuilding Corporation site.

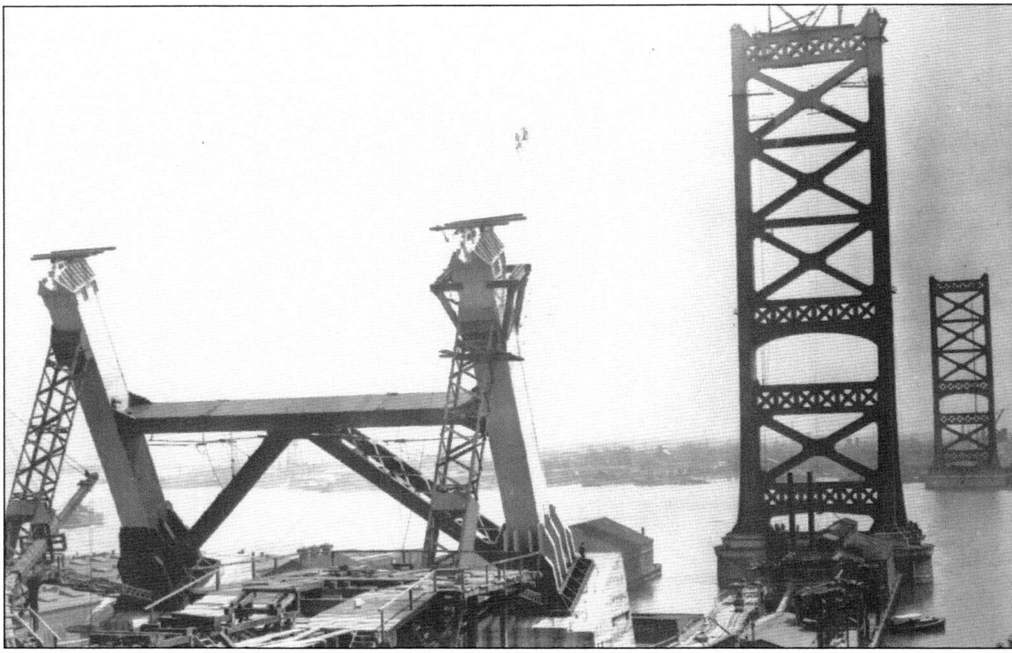

Work on the Philadelphia and Camden towers has been completed; the first phase of anchorage construction is also finished. In one month, footbridge construction will begin when the first 12 wires are strung out across the river and hoisted on top of the towers; these initial wires were used to create the footwalks that served as a working platform during the cable-stringing process. (Courtesy of Modjeski and Masters, Inc., Mechanicsburg, PA.)

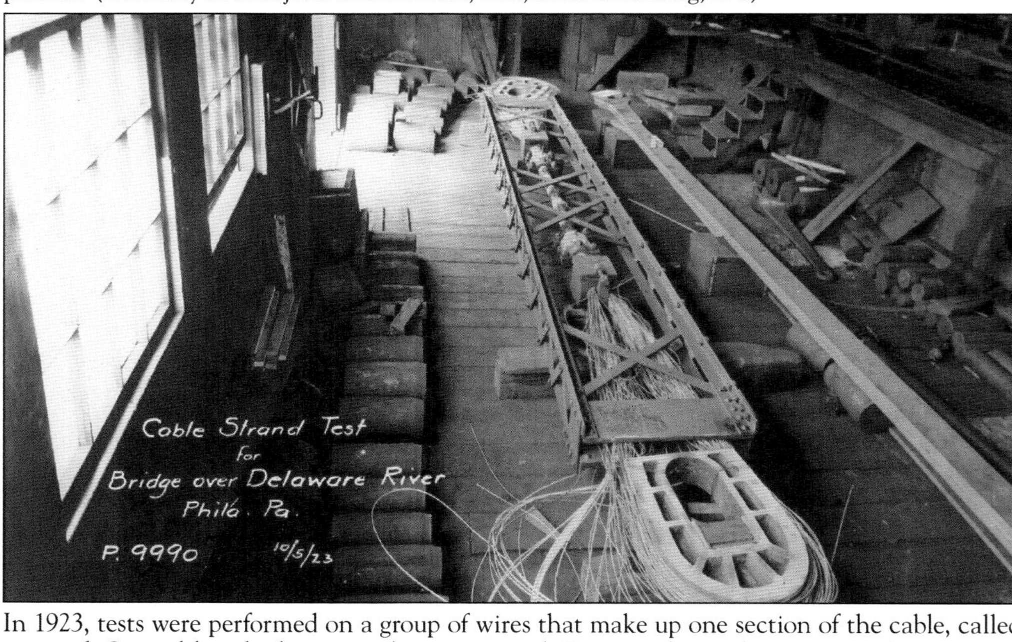

In 1923, tests were performed on a group of wires that make up one section of the cable, called a strand. Several hundred wires make up a strand. Dozens of strands are grouped together to form the cable. This picture shows one of the tests performed on a cable strand to determine the optimum strand strength based on the number of wires per strand.

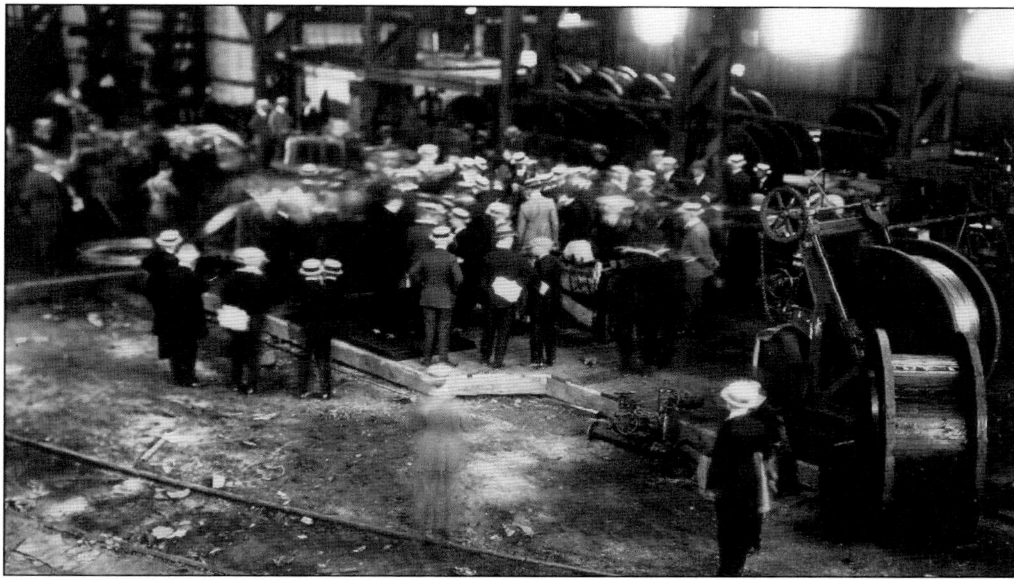

On June 10, 1924, ceremonies were held to commemorate the stringing of the first wires across the Delaware River. Dignitaries from Pennsylvania and New Jersey, along with commission members, inspected the cable-reeling plant in Camden and boarded three ships, the *J. Hampton Moore* (pictured at bottom), the *John Wanamaker*, and the *Ashbridge*. River traffic was halted as cranes hoisted 12 wires up to the tops of the towers. The wires had been pulled across the river over the previous five days. These wires were to support the footbridge used during cable stringing. After the wires were lifted to the top of the towers, the *John Wanamaker* darted beneath the bridge before any other boats, allowing the passengers to make the claim that they were the first to cross beneath the bridge.

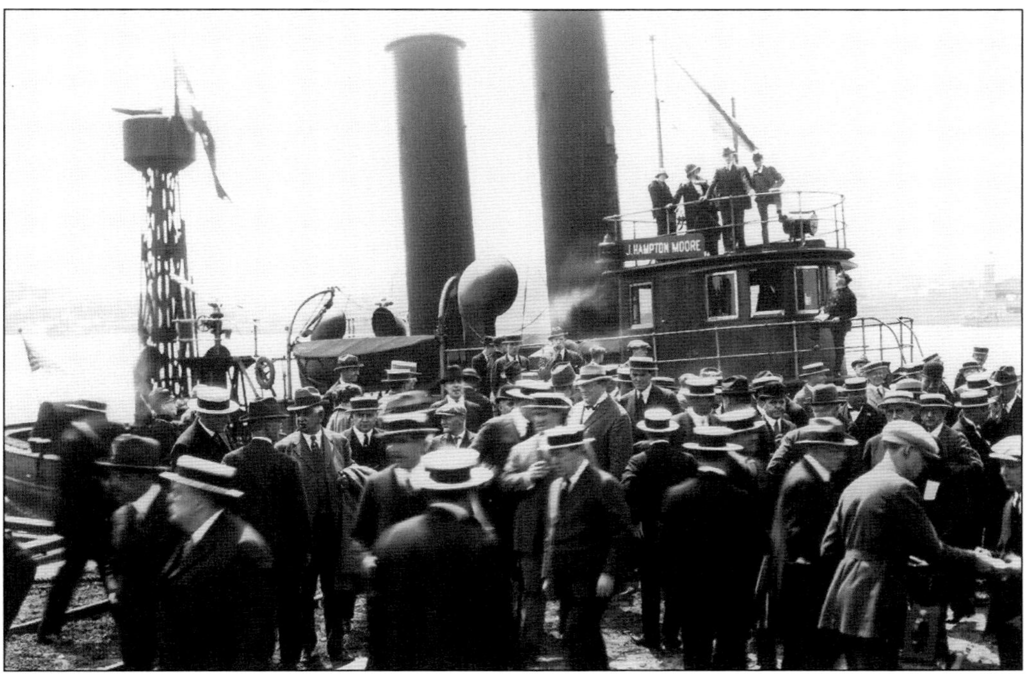

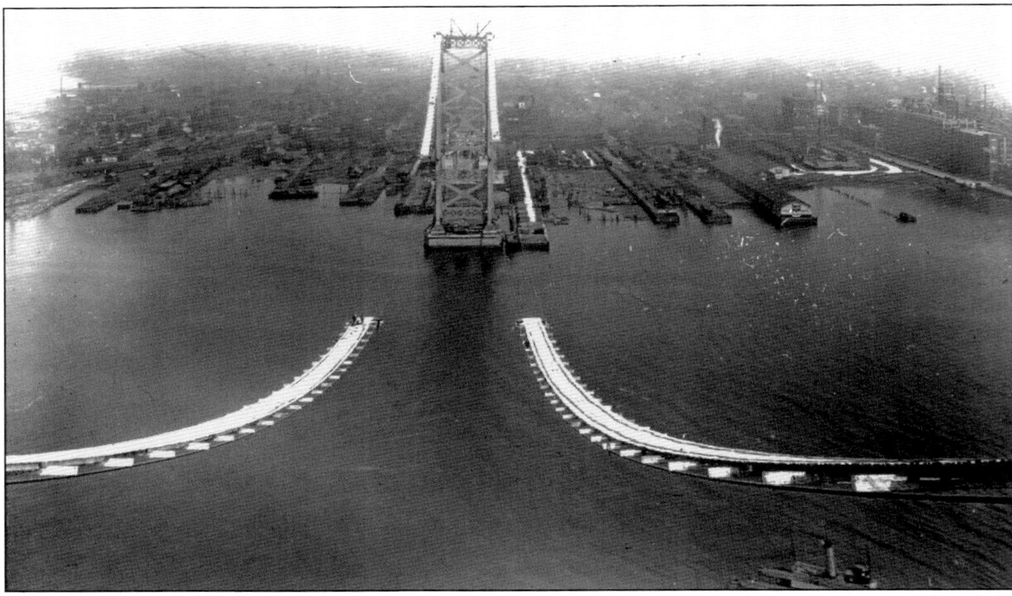

Six weeks after the initial 12 wires were hoisted into position, workers have nearly completed the footbridge construction. The footbridge was preassembled, hoisted into position, and placed on the wires. Each new section was fastened to its adjacent section, already set in place, by workmen stationed on a movable platform hung from the wires.

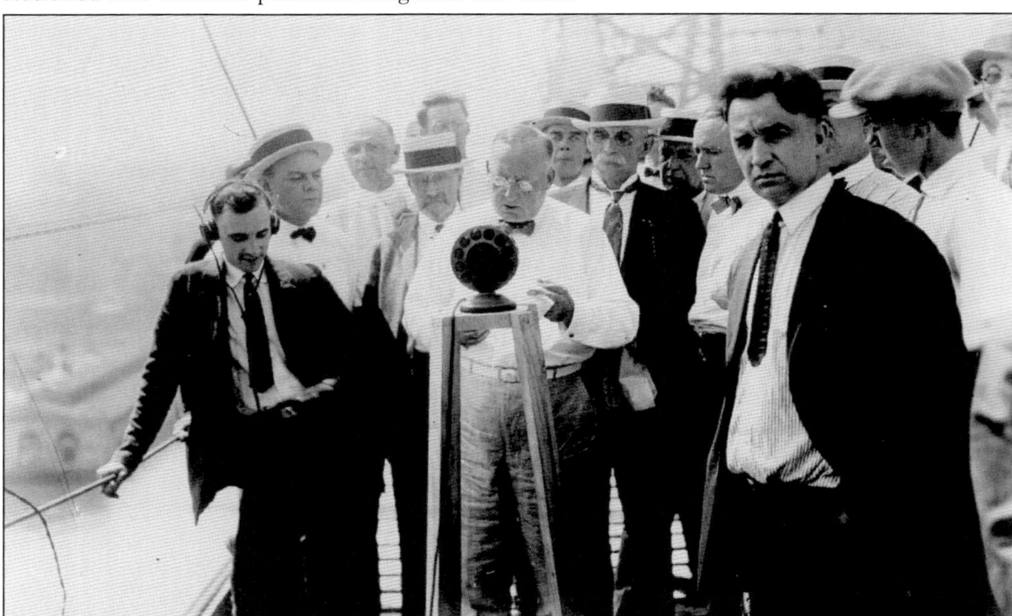

On August 8, 1924, the first official bridge crossing was made by members of the DRBJC and dignitaries from Pennsylvania and New Jersey. According to Charles Carswell, "Many guests of the commission who had been seen at previous inspections, and some officials, were 'unable to be present' at this first crossing," a daunting journey. At mid-span, local radio station WLIT set up equipment to broadcast the ceremony. Speeches were made by politicians, commission members, and Ralph Modjeski, who can be seen to the left of the man at the microphone.

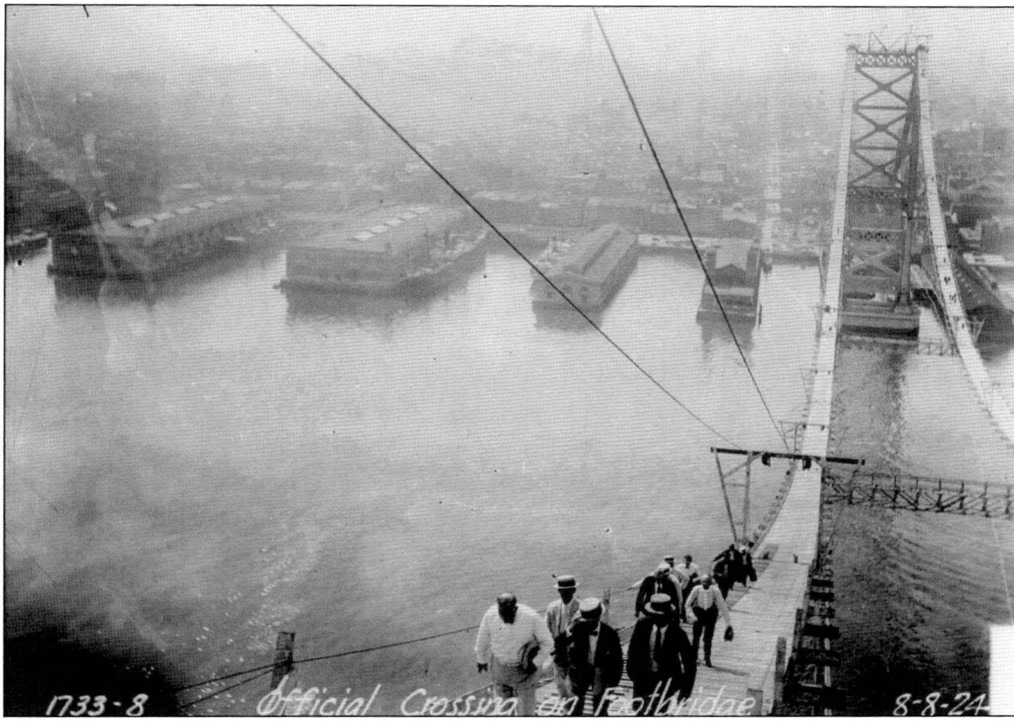

The official crossing, led by chief engineer Modjeski, began at the Philadelphia anchorage and proceeded east to Camden. Some of the guests making the trip walked the footbridge cautiously and continuously held onto the wire handrail. According to Carswell, "One of the guests unaccustomed to climbing such heights remarked that 'Washington had nothing on us.'"

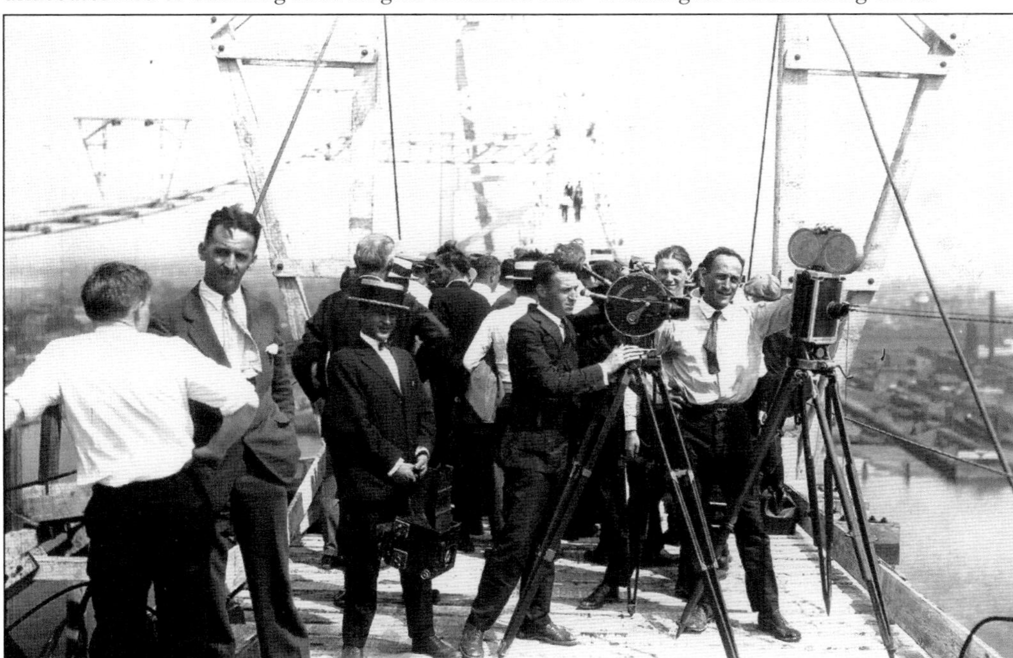

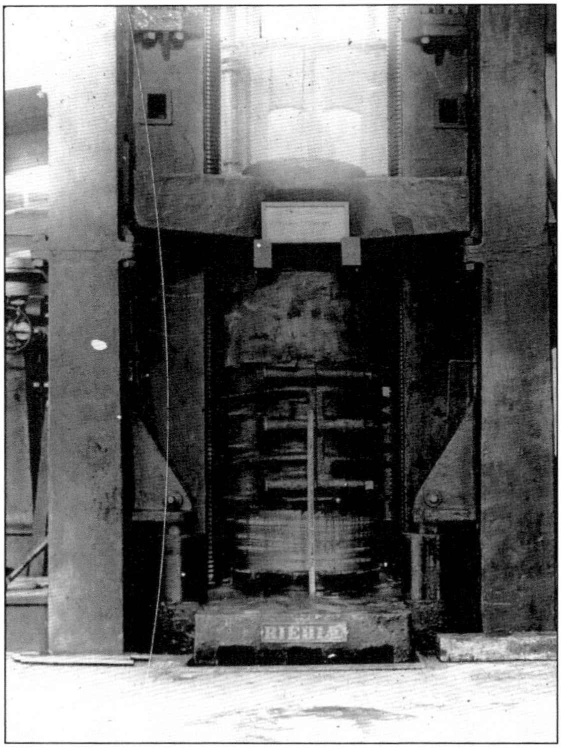

Before work on the main cable began, a five-foot-long model was created to test the strength of the cable and the ability of the equipment to properly compress it to its required shape and diameter. In the image at left, the model is undergoing a friction test on the cable band. The cable band is a steel casting used to secure suspender ropes that hold the roadway. This test is used to determine if the cable band can hold its position along the cable. The interior face of the band contains small cuts to keep it from slipping on the steeper sections of the cable. In the image below, the cable model is displayed in the courtyard of Philadelphia City Hall. The model was donated by the DRBJC to Philadelphia's Franklin Institute in the 1930s, where it was on display for more than 50 years.

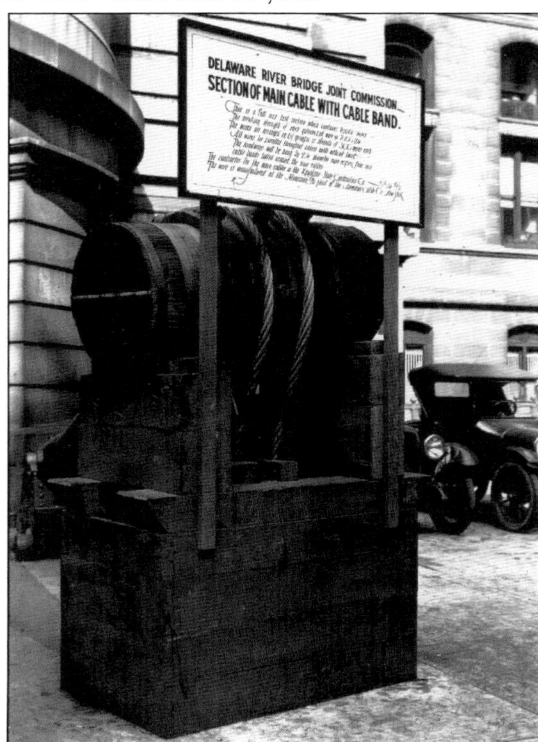

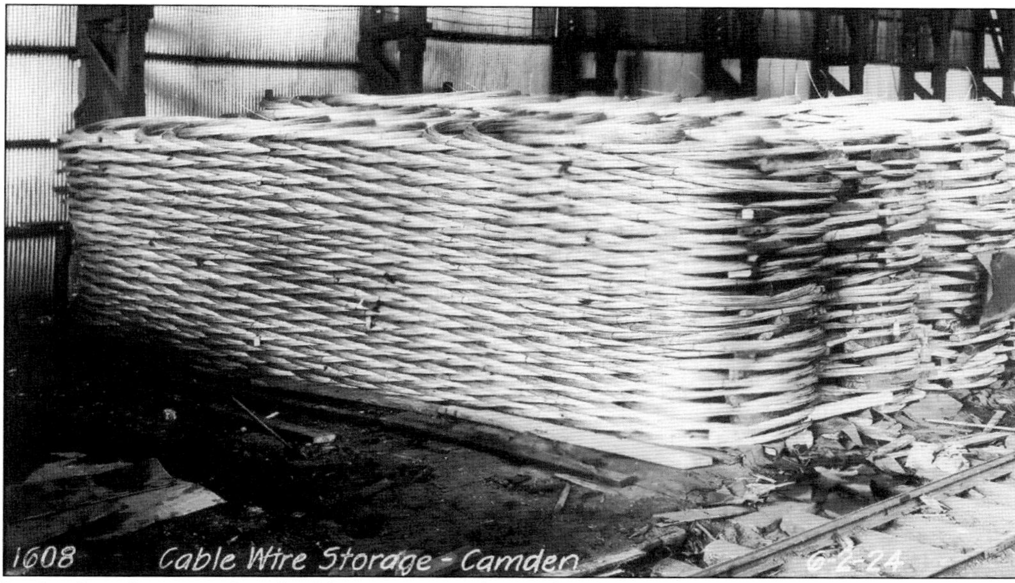

In the Camden wire plant building, wire was shipped to the site from American Cable Company's Monessen, Pennsylvania plant. The wires were shipped in coils of 3,300-foot lengths. The image above shows the cables after arrival from the Monessen plant. Once the cable arrived in Camden, it went through three phases designed to clean, strengthen, and protect the wire. The first phase involved the removal of mill scale (iron) from the wire. The second phase consisted of pulling the wire through a series of dies to draw it down to half of its original diameter. The third phase involved galvanizing the wire with zinc. After galvanization, the wire ends were spliced together to create a 100,000-foot continuous length and wound onto large reels. The image below shows the first reel of 100,000-foot wire.

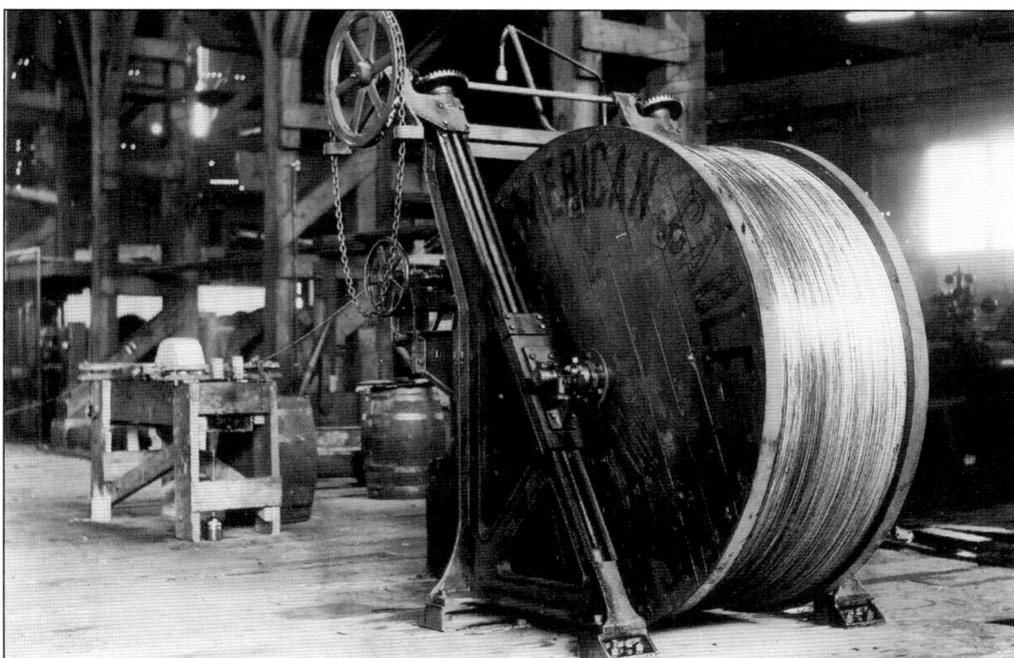

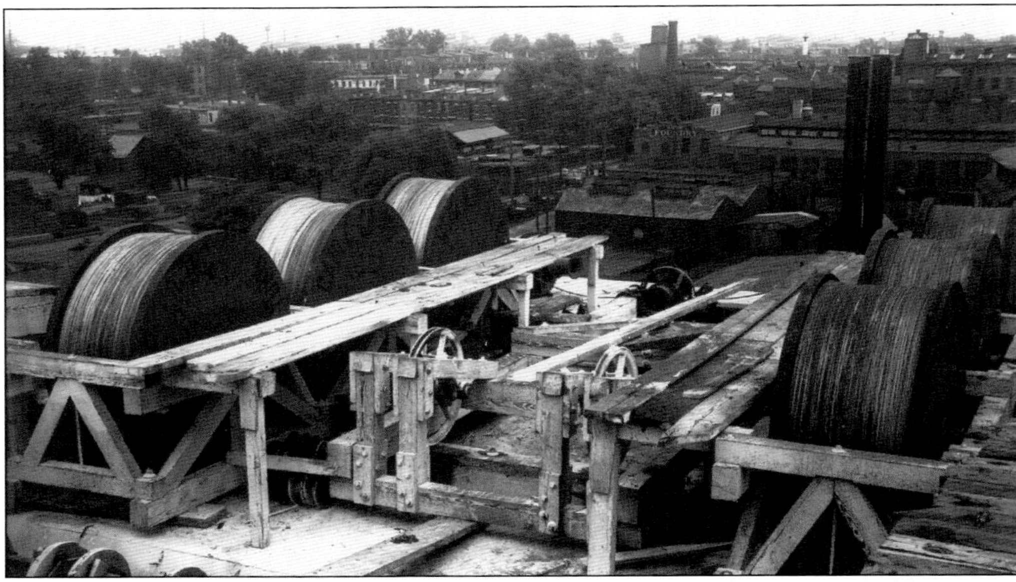

Six 100,000-foot cable reels are in place at the Camden anchorage. The process of cable stringing involved taking a loop of wire from a cable reel at the anchorage and wrapping it around a spinning wheel that was pulled across the river. Before the spinning wheel started its journey, one end of the wire was secured at the anchorage. As the spinning wheel moved across the river, the wire was pulled from the cable reel and plied out. With each wheel crossing, two wires were strung across the river.

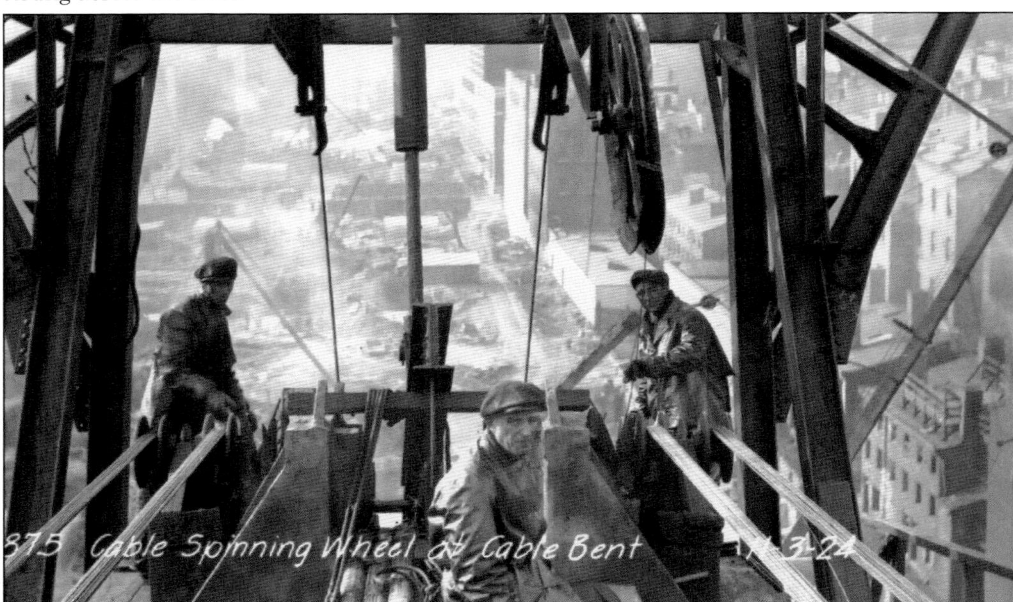

In November 1924, stringing of the main cable continues, with workers at the anchorage bent. The cable wheel seen above the worker's head on the right is pulling a loop of wire. The spinning wheel is attached to a continuous loop of wire called the "haulage rope," which pulls two spinning wheels attached at opposite ends; when one wheel was at the Philadelphia anchorage, the other was at the Camden anchorage. (Courtesy of Modjeski and Masters, Inc., Mechanicsburg, PA.)

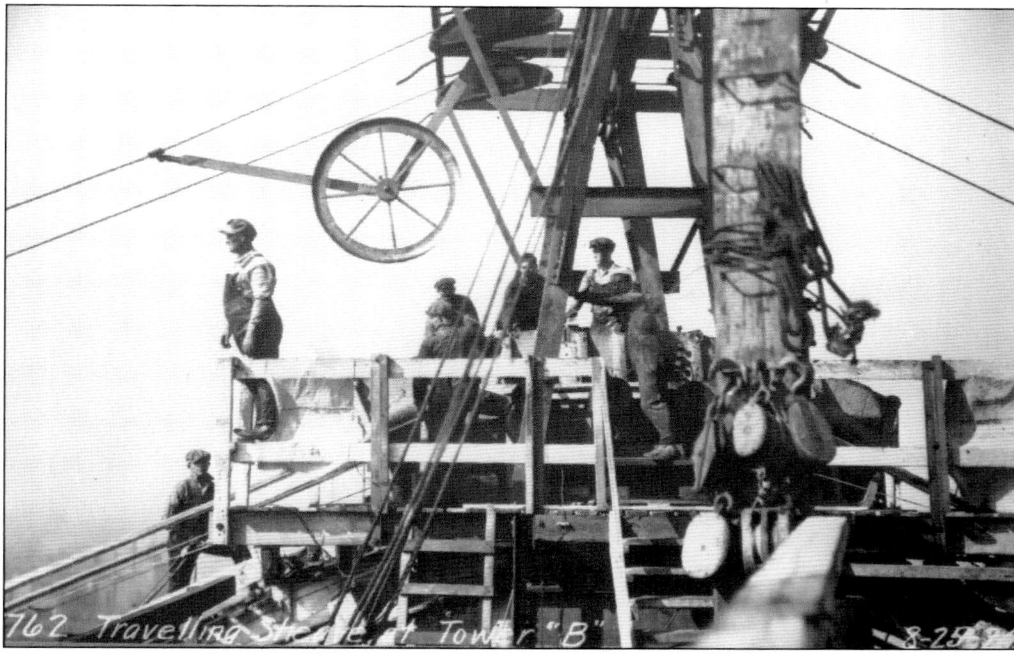

Seen in the image above, as the spinning wheel approaches the top of the Camden tower, workers take the two wires and adjust them into position within the wire strand. Cable stringing was performed 12 hours a day for 24 weeks. The most wires strung on a single day was 578, resulting in an average crossing time of 10 minutes at an average speed of four miles per hour for each spinning wheel. When the wheel arrived at both anchorages, workers removed the wire loop and positioned it around a device called a strand shoe, which secured the wire to the eyebars within the anchorage. After the loop was removed, a new loop of wire was wrapped around the wheel and sent across the river. In the image below, workers pose as the two spinning wheels pass each other at mid-span. Spaced evenly across both footbrigdes, timber frames were constructed to support the haulage rope. (Above, courtesy of Modjeski and Masters, Inc., Mechanicsburg, PA.)

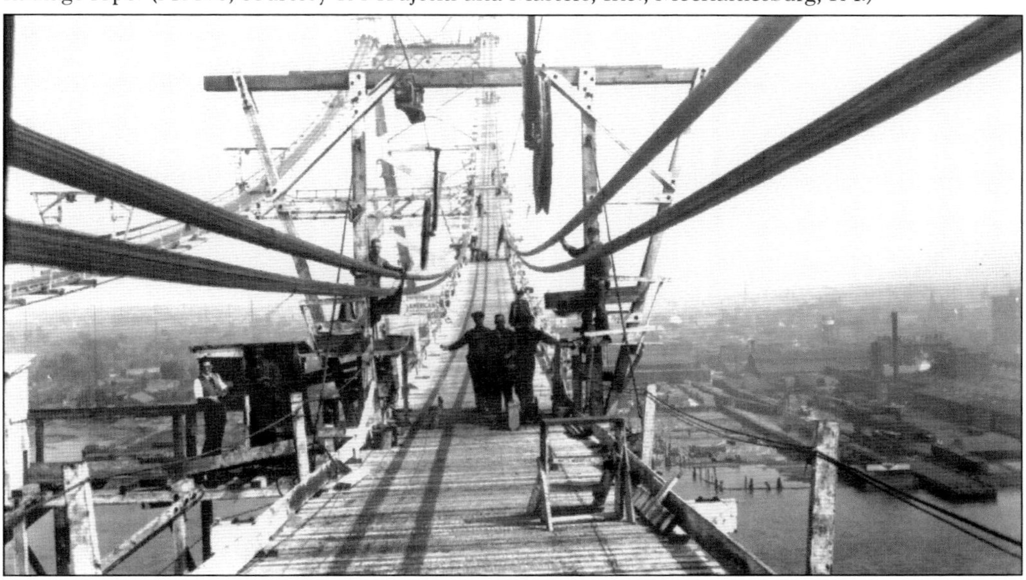

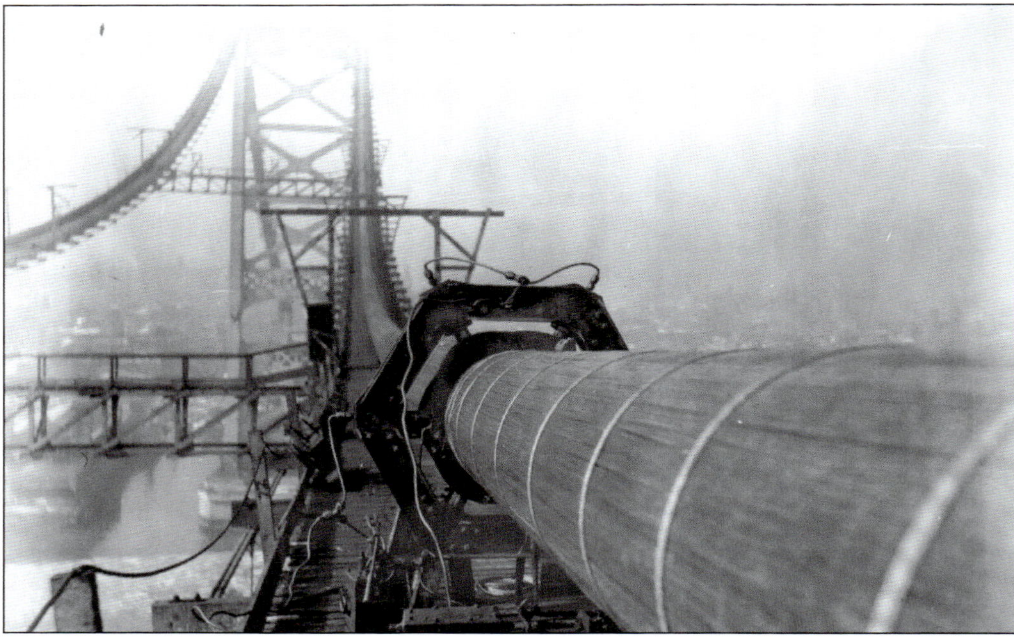

After the cable-stringing process was finished, the numerous cable strands in the main cable formed a hexagonal shape. To create the final circular shape, a compacting machine was created that squeezed the cable with six hydraulic jacks connected as a single unit. Each jack applied a 30-ton load. Compaction along the length of the cable took two months.

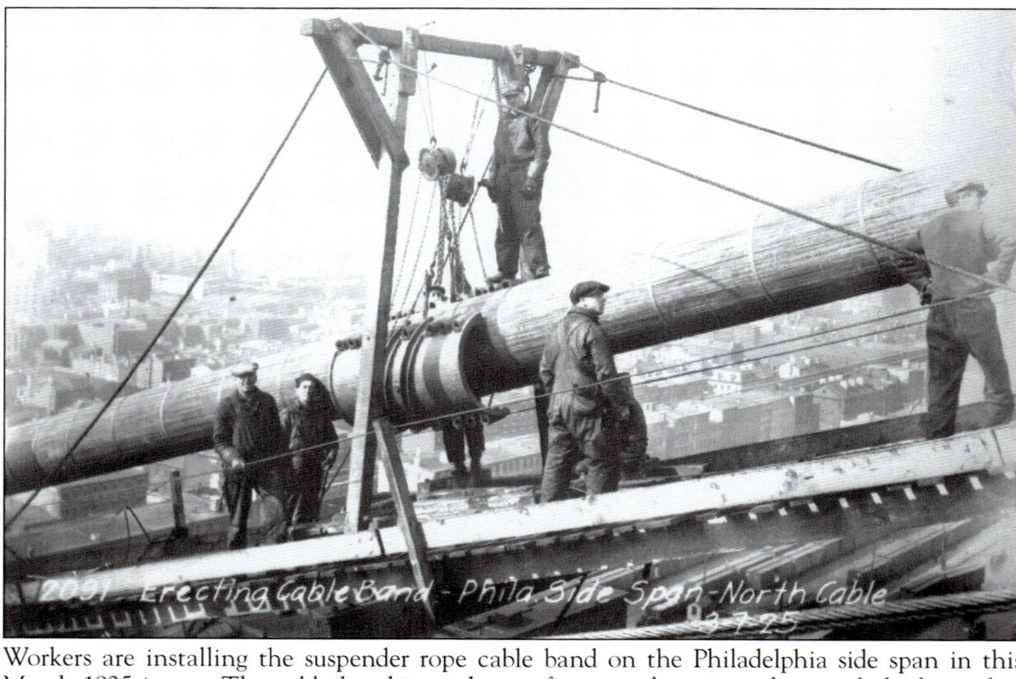

Workers are installing the suspender rope cable band on the Philadelphia side span in this March 1925 image. The cable band is made up of two steel castings that are bolted together around the main cable. The band allows the suspender rope to be draped around the main cable through ribs attached to the exterior of the casting.

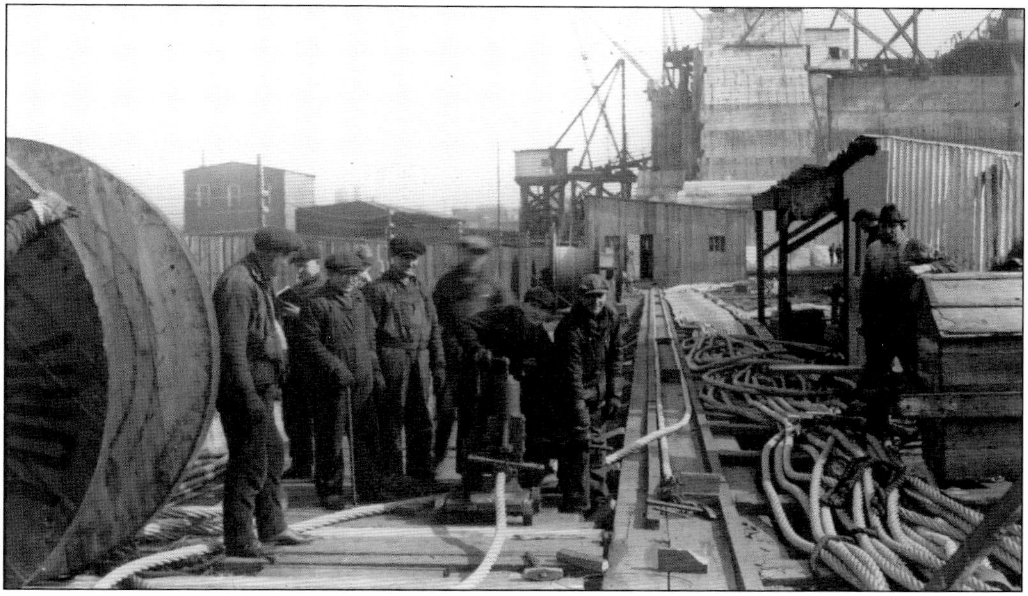

Prior to hanging, the individual suspender ropes on the main cable had to be measured to their required length based on their location along the cable. Suspender ropes at mid-span were considerably shorter than those at the towers. Workers in the photograph above are measuring and cutting the suspender ropes to the proper lengths. As seen in the image below, after the suspender ropes were cut to length, the ends of each rope were opened, and the individual wires were separated for cleaning. The cleaning involved removing the lubricant that had been applied within the ropes during the manufacturing process. After cleaning, the ends of the ropes were attached to a steel socket that was used to secure a section of the roadway's stiffening truss to the suspender rope. To create a secure grip between the socket and wires, zinc was poured into the socket and around the individual wires.

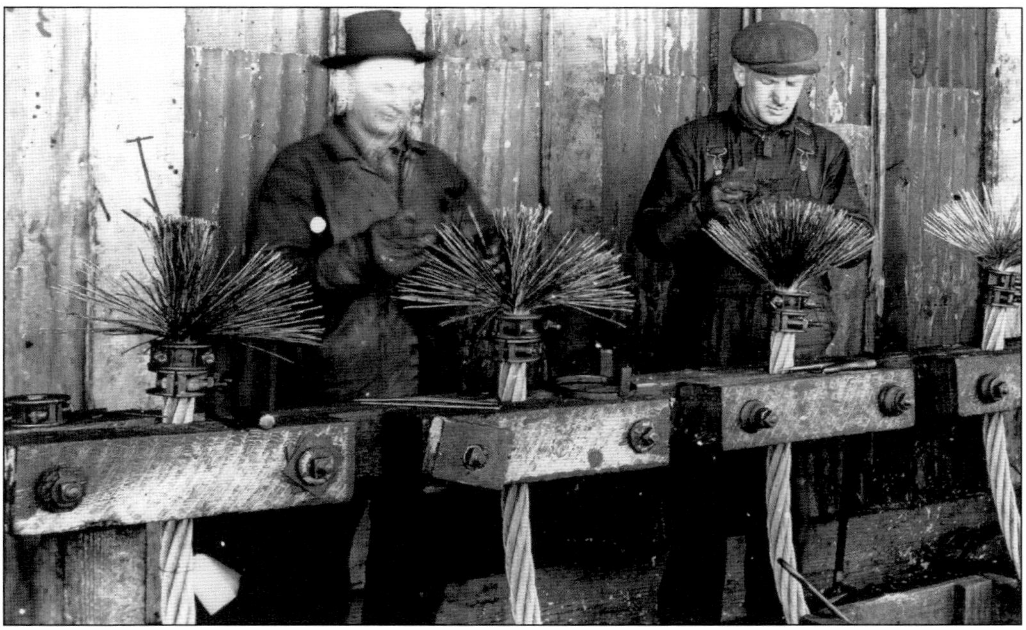

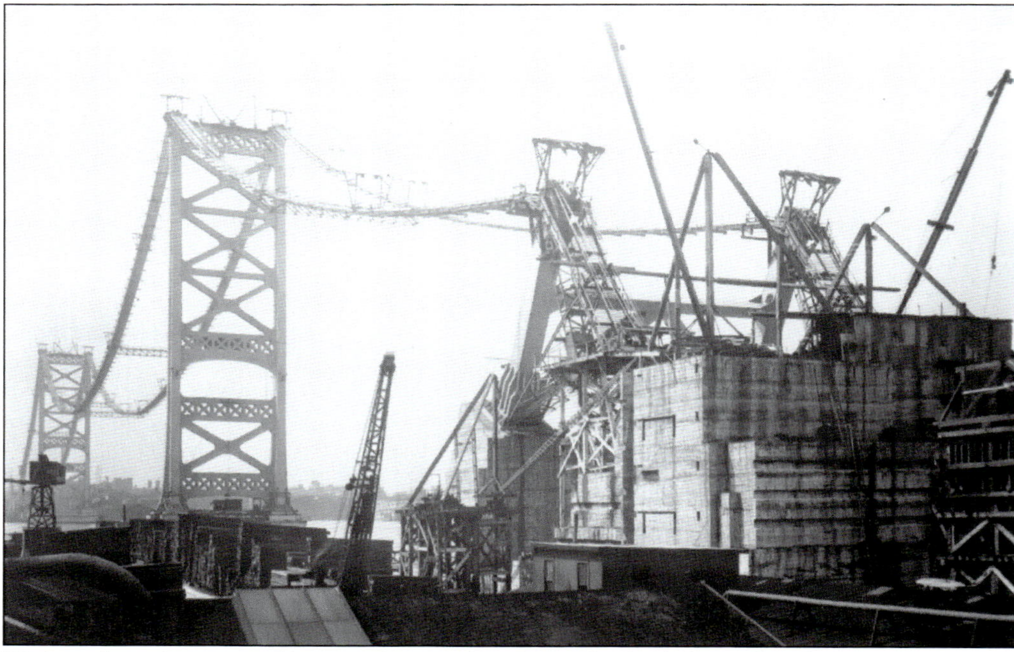

With work on the main cables complete, workers begin the final phase of anchorage construction: installing granite blocks around the base of the Camden anchorage. The granite blocks were quarried from Stone Mountain, the same location where the tower pier granite was quarried. Work also proceeded on the elevator towers on the north and south sides of each anchorage.

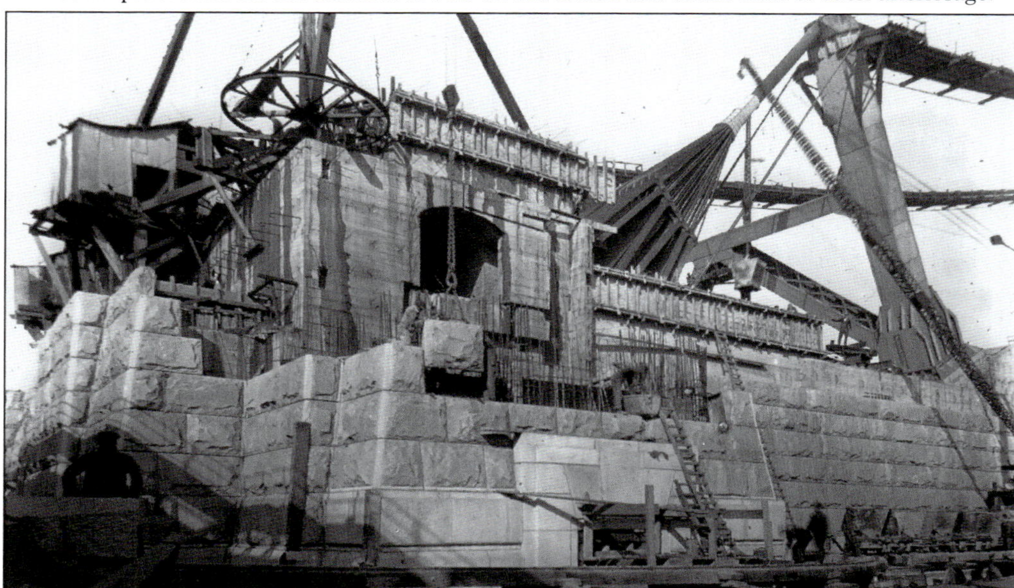

Work continues on the granite block installation on the exterior face of the Philadelphia anchorage. The opening seen in the center is a passageway that was to be used to transfer between the Frankford Elevated (its shadow can be seen in the lower left corner) and the bridge's rapid-transit line. This corridor extends to the north side of the anchorage. A perpendicular corridor connects the passage with the west face of the anchorage.

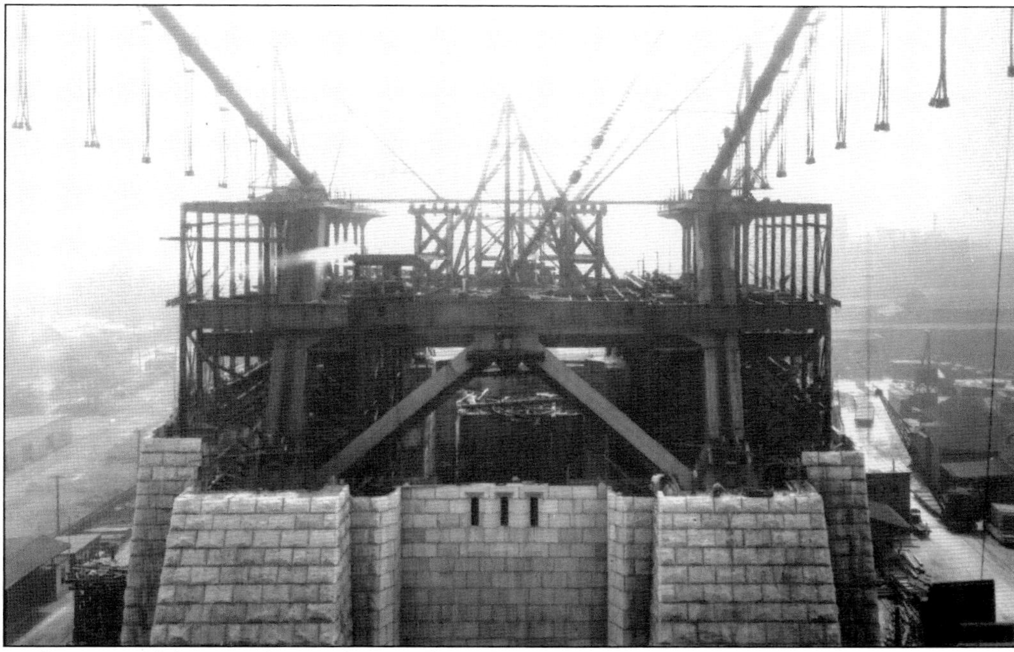

In this view of the Camden anchorage from the adjacent tower, several pairs of suspender ropes have been installed along the main cable close to the anchorage. Workers have also constructed the steel skeleton used to support the roadway, rapid-transit track areas, and the pedestrian walkways. (Courtesy of Modjeski and Masters, Inc., Mechanicsburg, PA.)

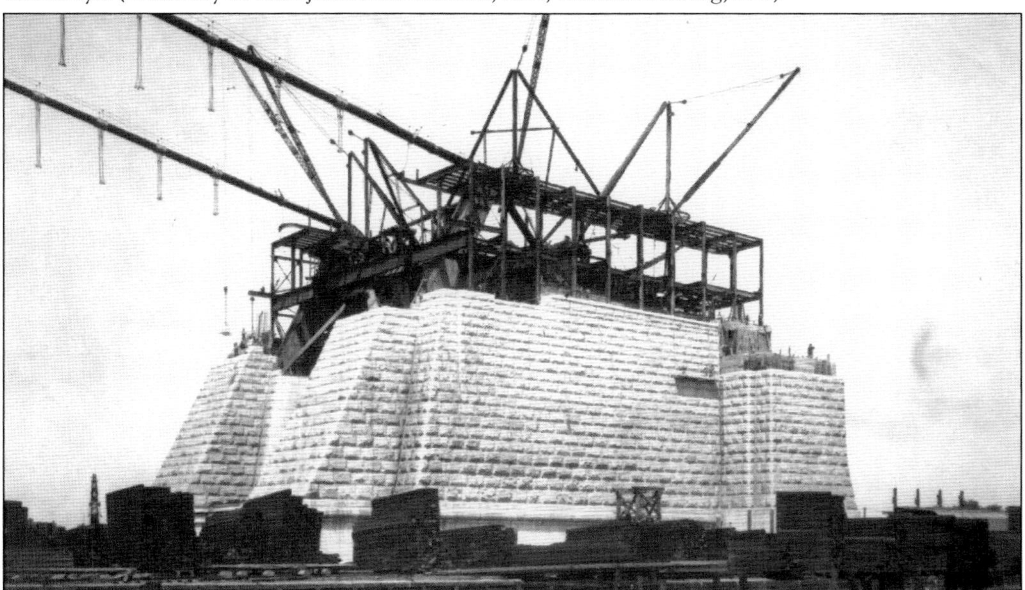

As the steel framing for the roadway along the suspended spans nears completion, this image of the Camden anchorage shows the progress of the granite installation for the exterior face. The anchorage's steel skeleton is quickly disappearing as the concrete core and granite facade reach the roadway level. Construction will continue as workers build the two anchorage towers. (Courtesy of Modjeski and Masters, Inc., Mechanicsburg, PA.)

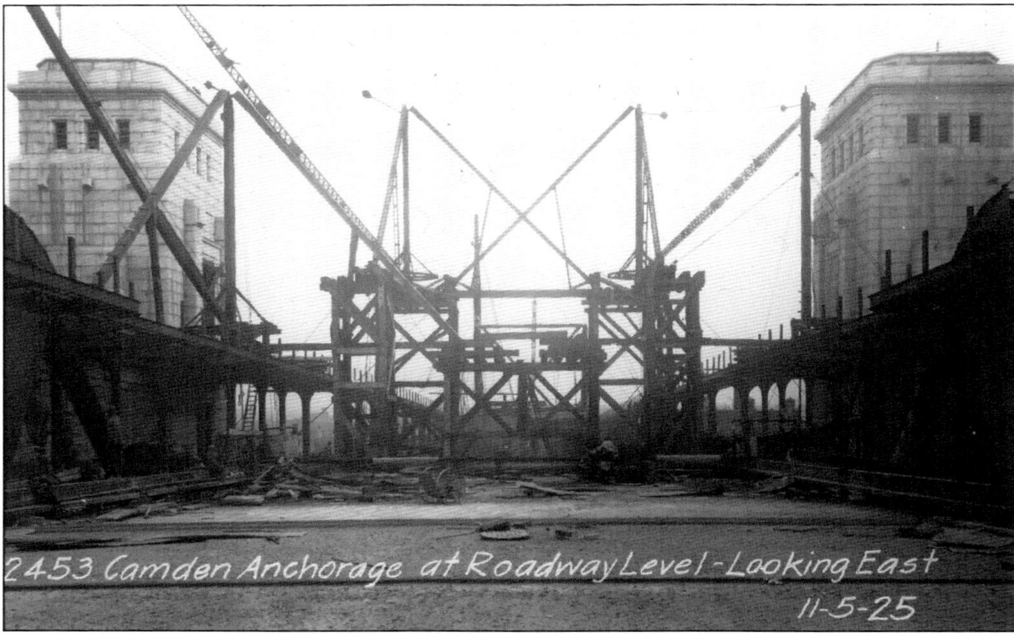

Work on the anchorage towers is complete. The two towers were designed to serve as elevator shafts for those wanting to transfer between the bridge walkway, the rapid-transit station, street level, and the Frankford Elevated in Philadelphia. Elevator lobbies at each public level featured ornamental tiles created by the Enfield Pottery and Tile Works in Springfield, Pennsylvania.

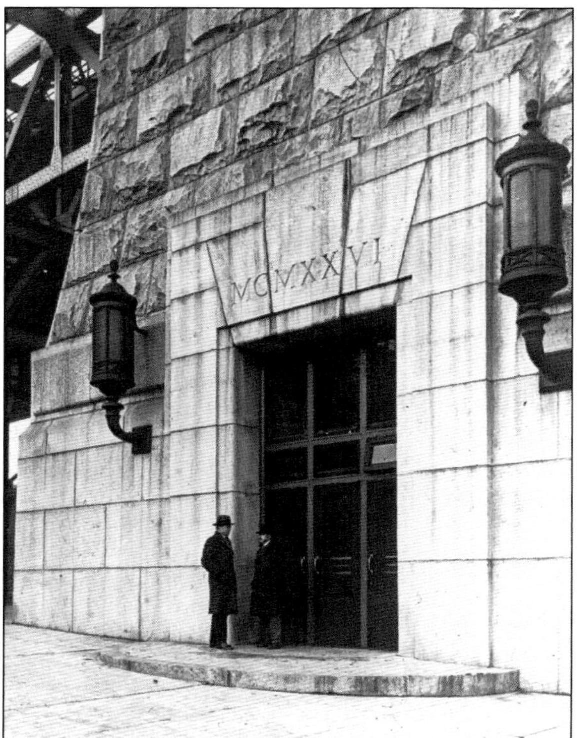

At the anchorage's street-level entrance, two individuals stand next to doors that would have led pedestrians into the elevator lobby. Although the elevators were installed, they were never opened to public use.

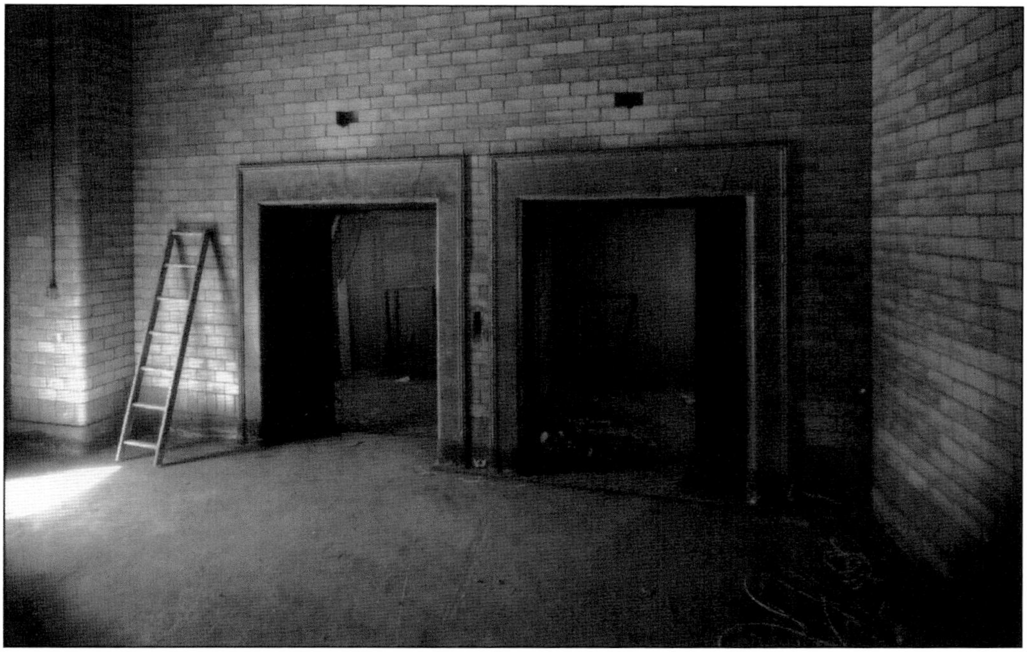

This image shows the south lobby area at the street level of the Philadelphia anchorage. Pedestrians would have entered the anchorage from a doorway seen at left, taking one of three elevators up to either the walkway level, the station crossover level, or to an interchange with the Frankford Elevated Railroad. The image, taken in 2000, shows two of the elevators ready for passengers who will never arrive. (Courtesy of James B. Abbott, copyright 2000.)

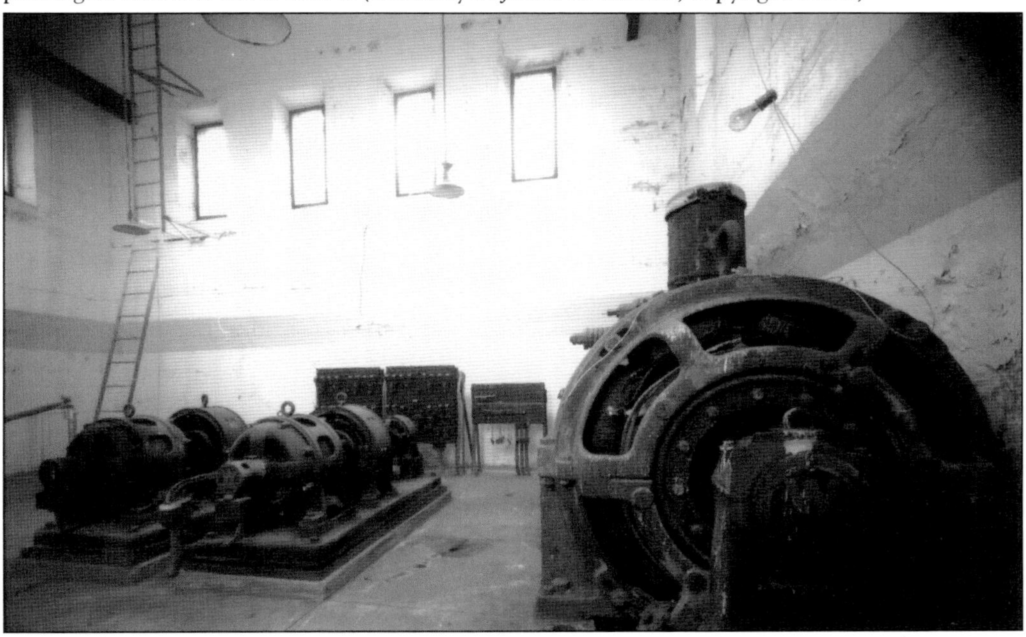

Motors, still housed in the anchorage towers, were to lift the two elevators seen in the previous image. This room in the anchorage towers is now illuminated at night as part of the bridge's decorative lighting. (Courtesy of James B. Abbott, copyright 2000.)

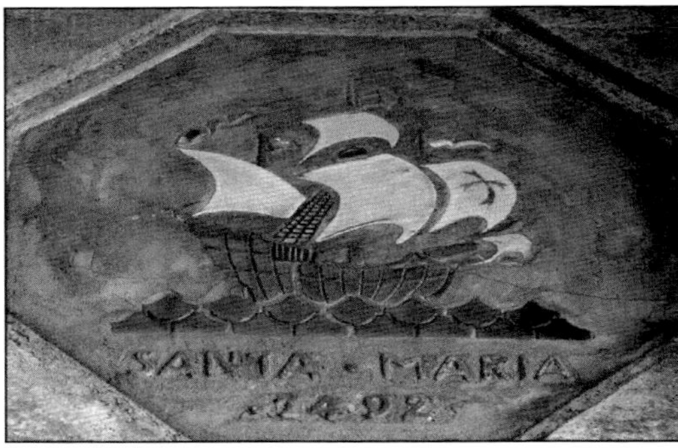

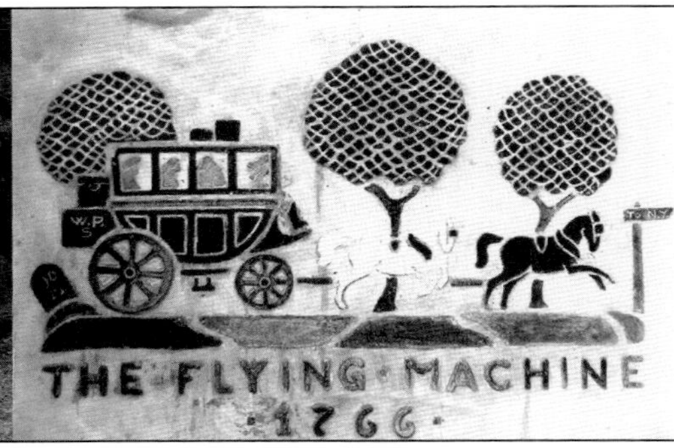

In the lobby area at the walkway level, decorative tiles created by the Enfield Pottery and Tile Works are "set in the walls depicting milestones of development of transportation in America," as noted by Charles Carswell. The tiles seen here depict the *Santa Maria*; the "Flying Machine" (an improved stage coach that reduced the travel time between Philadelphia and New York to two days, or three days in winter); Robert Fulton's steamboat, the *Clermont*; and the *Constitution* sailing ship.

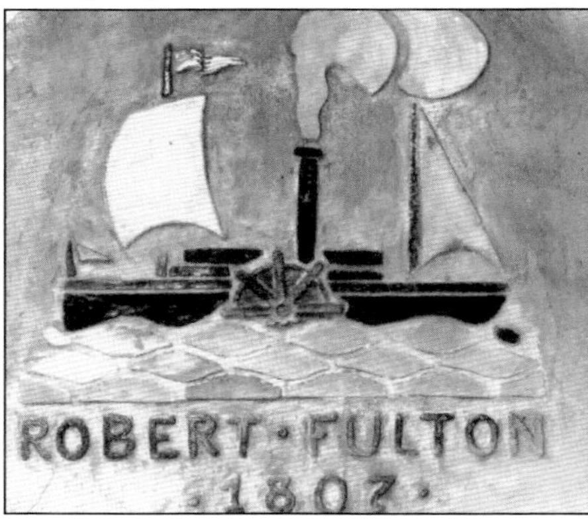

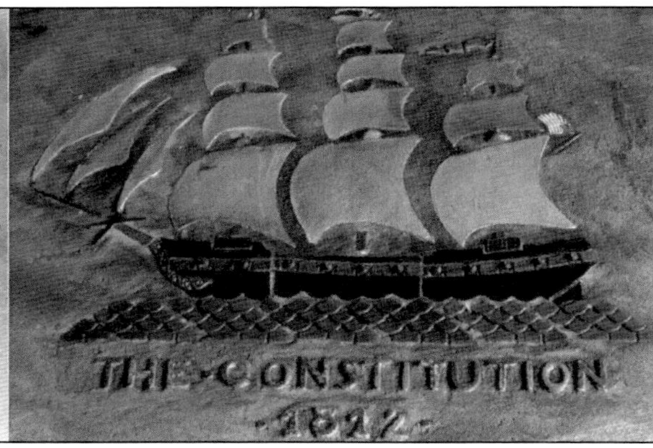

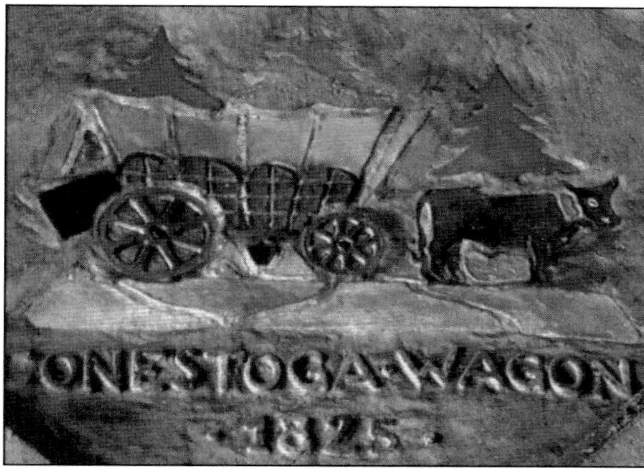

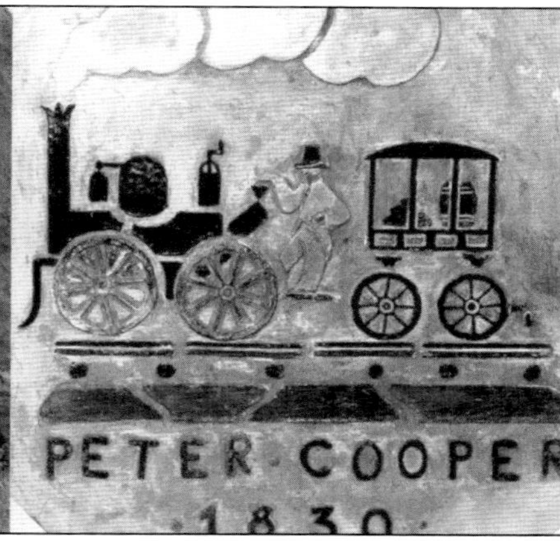

Pictured here are tiles depicting the Conestoga wagon, the "Peter Cooper" (commemorating rail travel), and the airship *Shenandoah.* The *Shenandoah* crashed in September 1925, not long before the tiles were set in place. As expressed by Walter S. Andariese in his book on the bridge's history, the workers setting this particular tile "must have had misgivings."

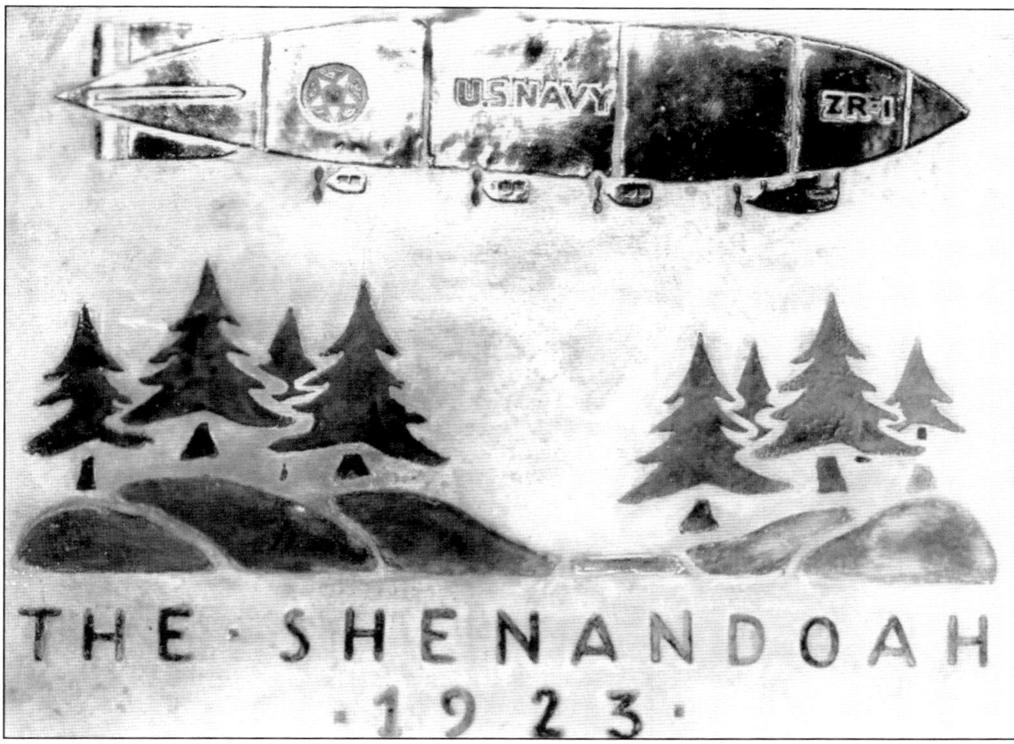

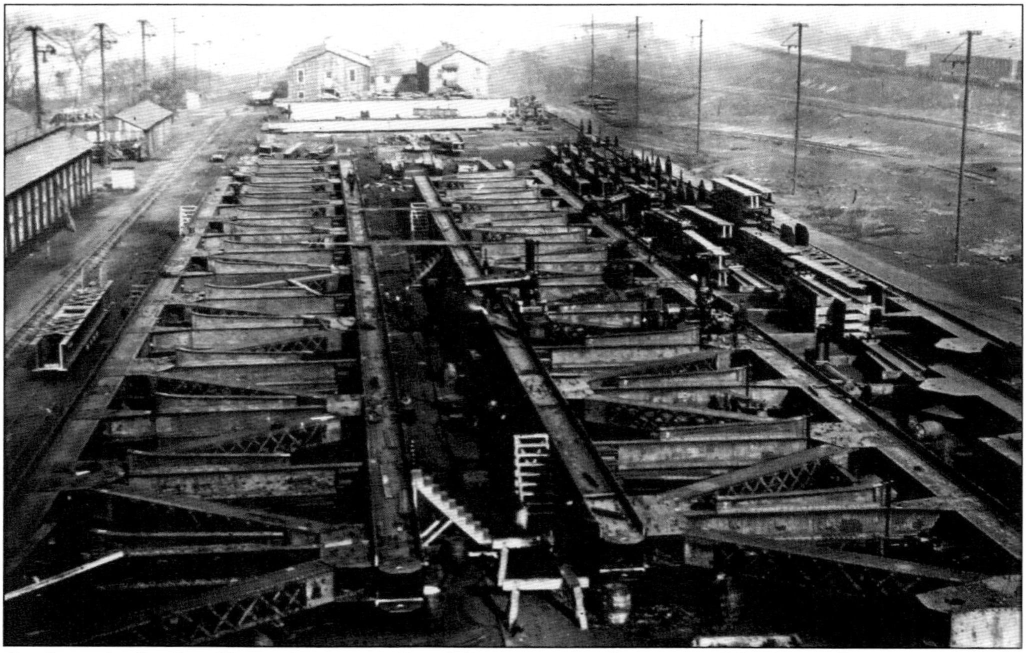

At their Ambridge Shop, American Steel Company workers assemble the stiffening trusses for the main span. As with other parts of the bridge, sections were preassembled off-site to ensure a proper fit. The sections were disassembled into smaller sections, then shipped to the construction site for final installation.

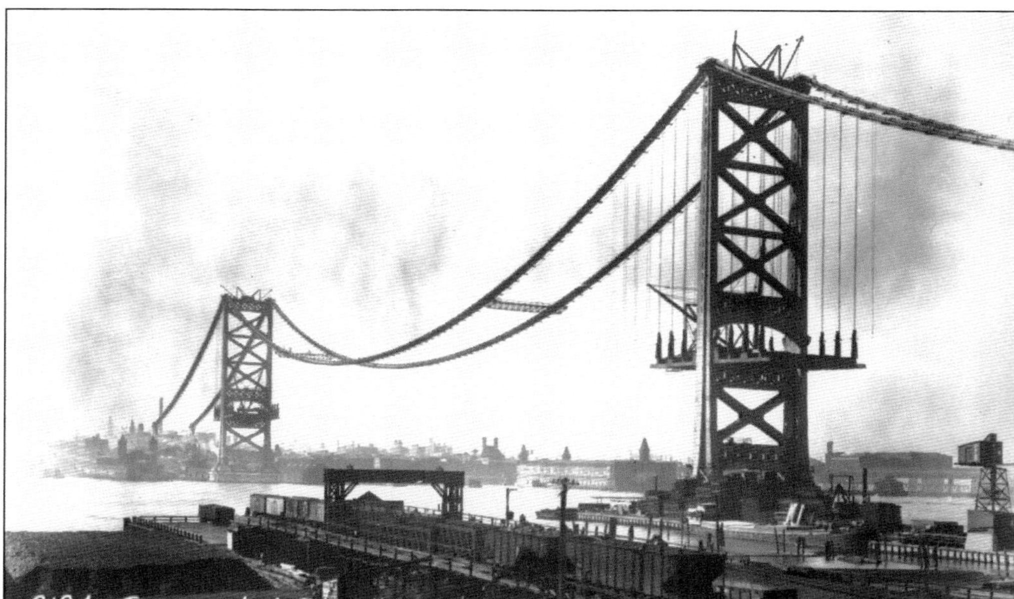

Installation of the suspended spans began in March 1925. To reduce stress on the cable, the roadway sections were first installed adjacent to the towers. The load on the cables is distributed evenly as the roadway sections are installed from the towers to mid-span and toward both anchorages. Cranes mounted on the tower installed the initial roadway sections.

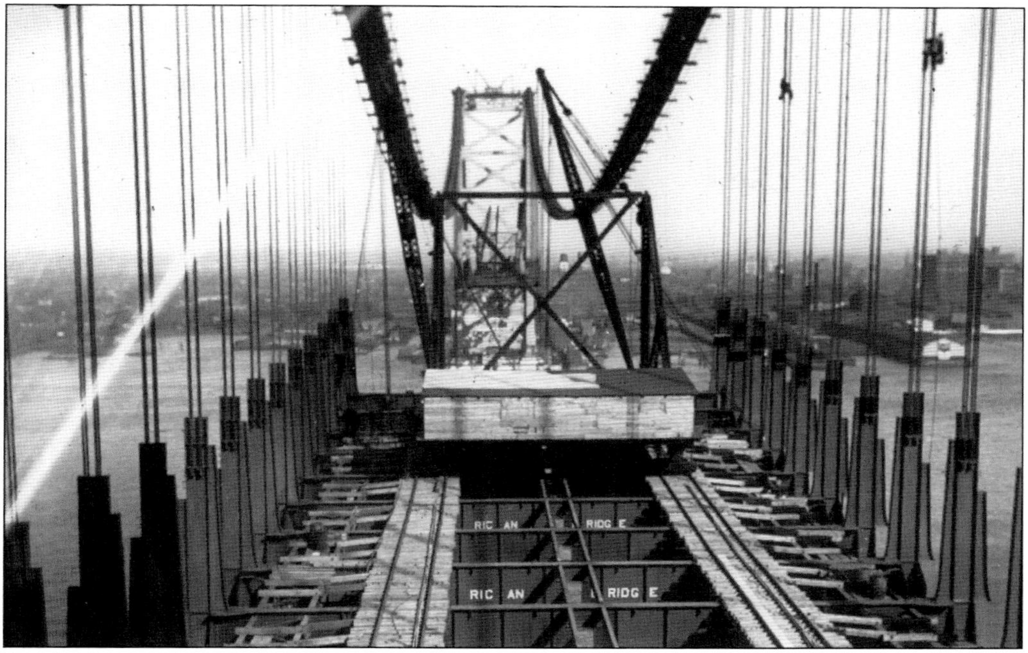

Roadway construction travelers move along the bridge on temporary tracks to hoist roadway sections from barges beneath the bridge. This manner of installation proved successful, except where the bridge crosses Delaware Avenue adjacent to the Philadelphia anchorage. To complete the installation across Delaware Avenue, the steel was delivered to the site by truck and hoisted into place.

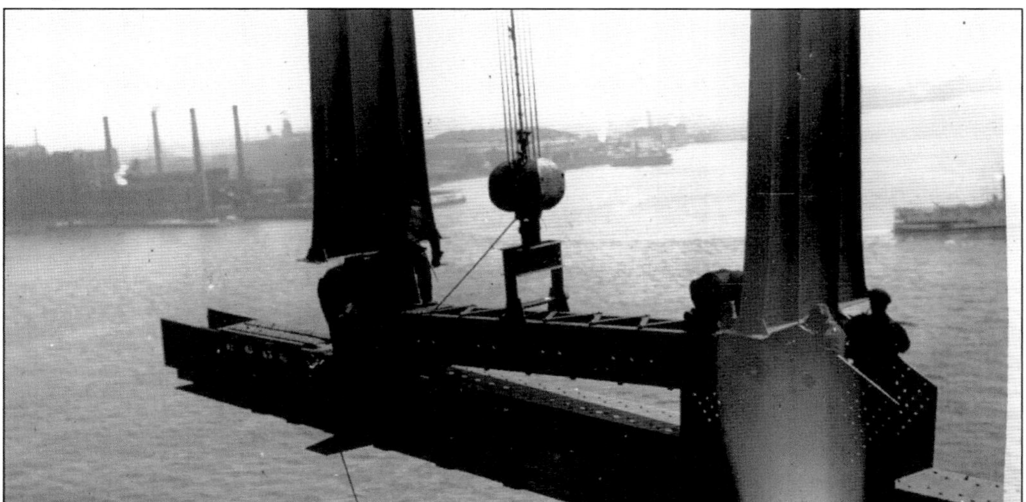

On the Philadelphia portion of the main span, workers lift the lower chord into position. Once in position, workers will secure it to the adjacent gusset plate. Steel installation along the roadway progressed through several stages to keep the weight uniformly distributed throughout the main cable. The initial stage involved the installation of the vertical members, the lower chords, and floor beams. The top chords, diagonal members, floor stringers, and braces were installed during subsequent stages.

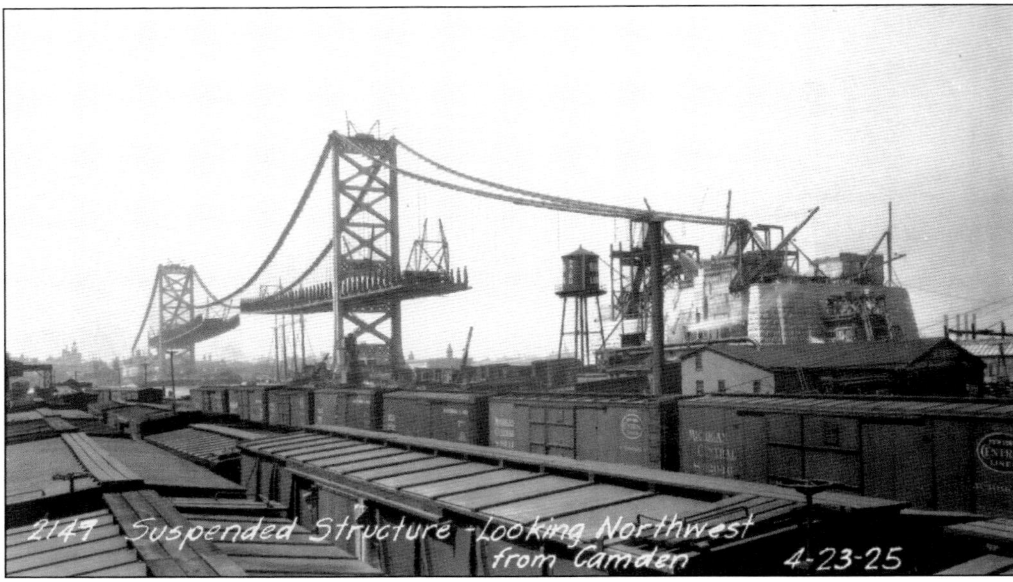

In the image above, taken toward the end of April 1925, steel erection for the main span is nearly complete, whereas steel erection on the side spans has not reached the halfway point. Charles Carswell noted that "delivery of suspension steel favored the main span over the side spans at one point [during construction]. The added steel on the main span caused the towers to bend 18 inches toward the center of the river." Newspapers articles noted this "bending," which required bridge engineers to explain that the towers were performing as intended; the bending was within the 20-inch design tolerance for the towers. The photograph below shows that the travelers have almost closed the gap on the main span steel erection. To assist with closing the gap, workers have also begun to install the top chord steel near mid-span. Footbridges that were used during the cable-stringing process are being dismantled. (Courtesy of Modjeski and Masters, Inc., Mechanicsburg, PA.)

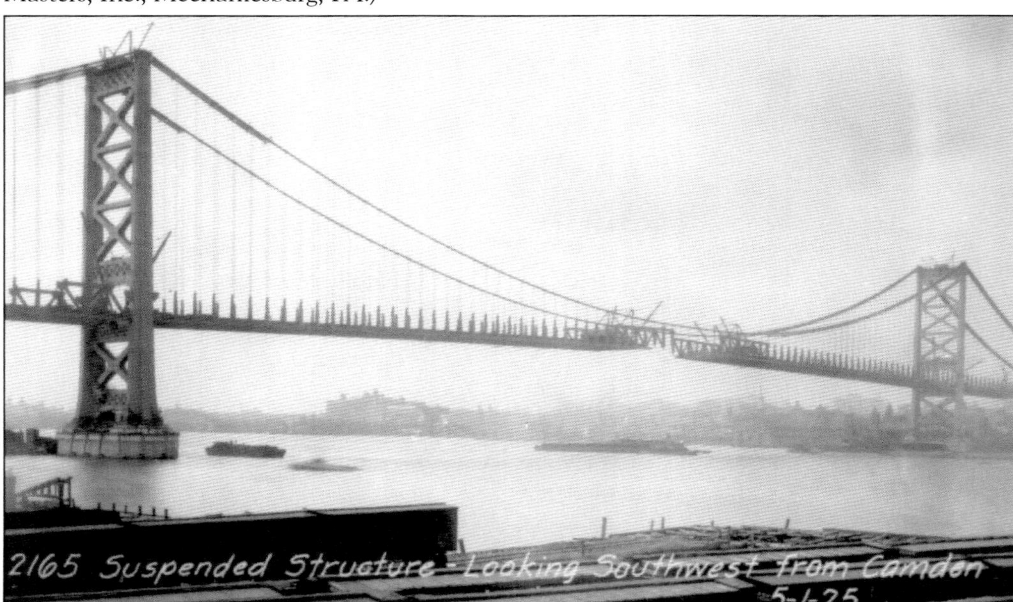

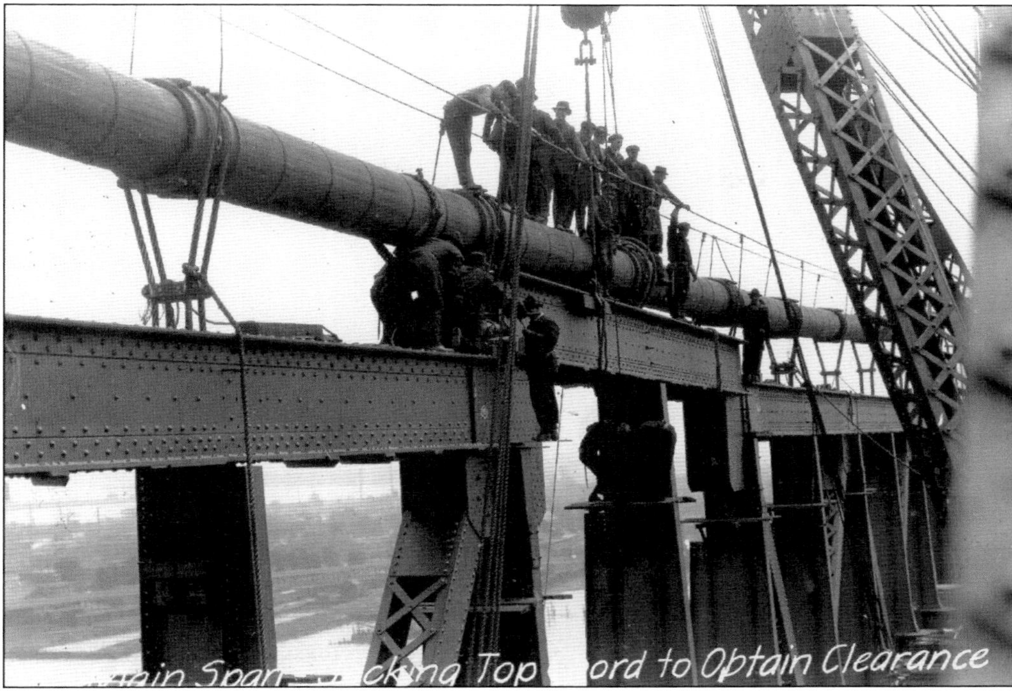

On May 7, 1925, workers are seen jacking the top chord into position. Installation of this section brought closure to the main span. Since the steel for the main and side spans was not delivered at the same rate, the side spans were completed at the anchorages in July 1925. As steel was installed on the side spans, the towers returned to their vertical position.

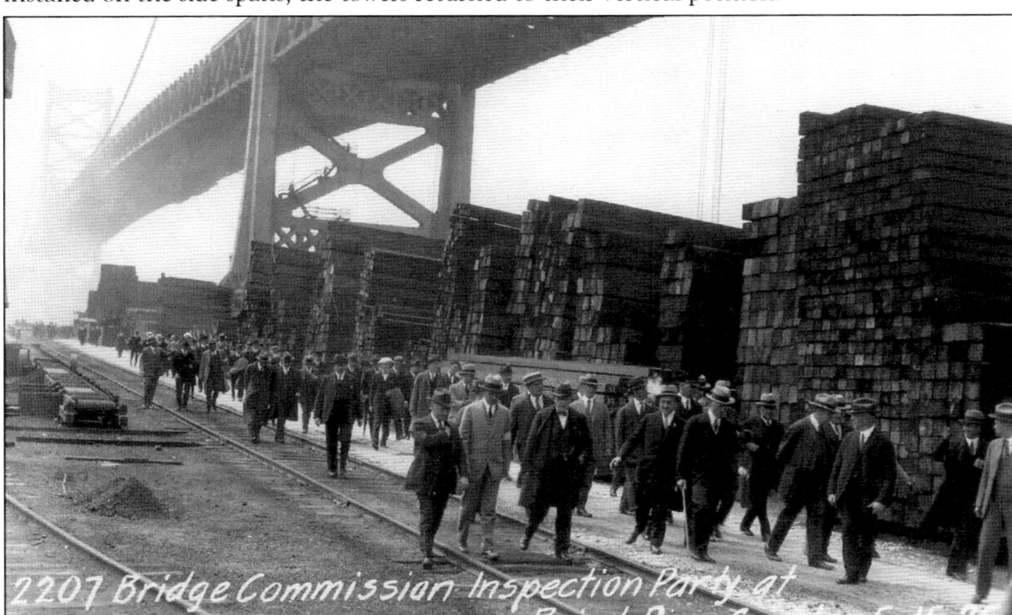

In May 1925, an inspection party consisting of commissioners and dignitaries from both states made a site visit following completion of the main span's steel deck.

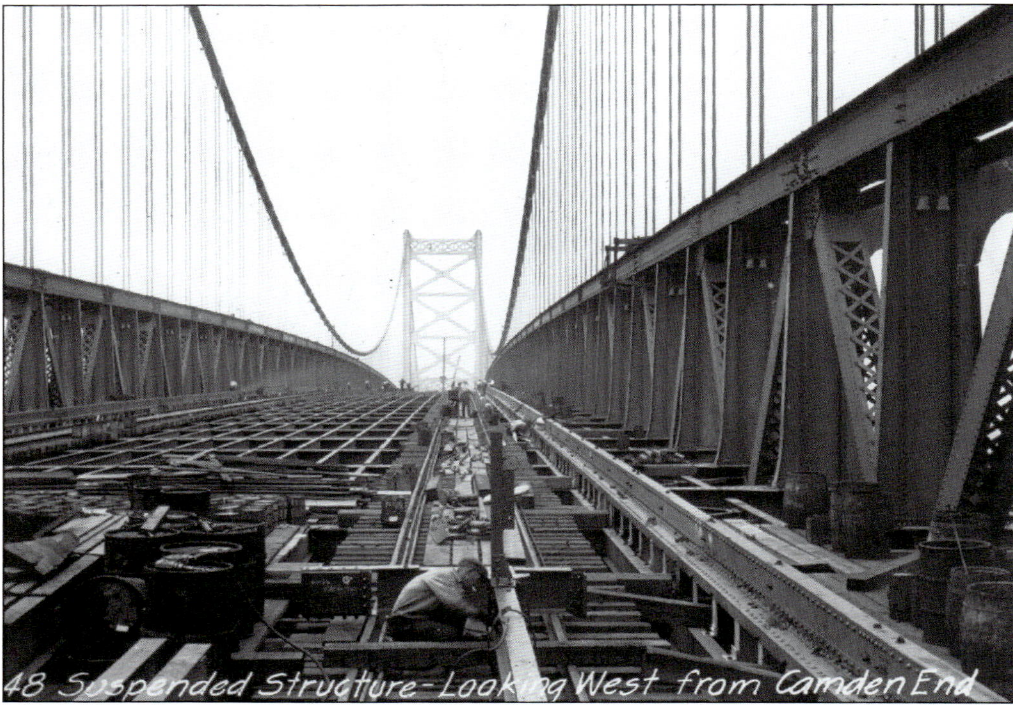

At the end of July 1925, workers could make use of the temporary track that the traveler cranes operated on as a platform. The individual in the foreground is working on a steel member for the pedestrian walkway. The vertical post attached to the steel member will be used to secure decorative cast iron panels that line the exterior of the walkway.

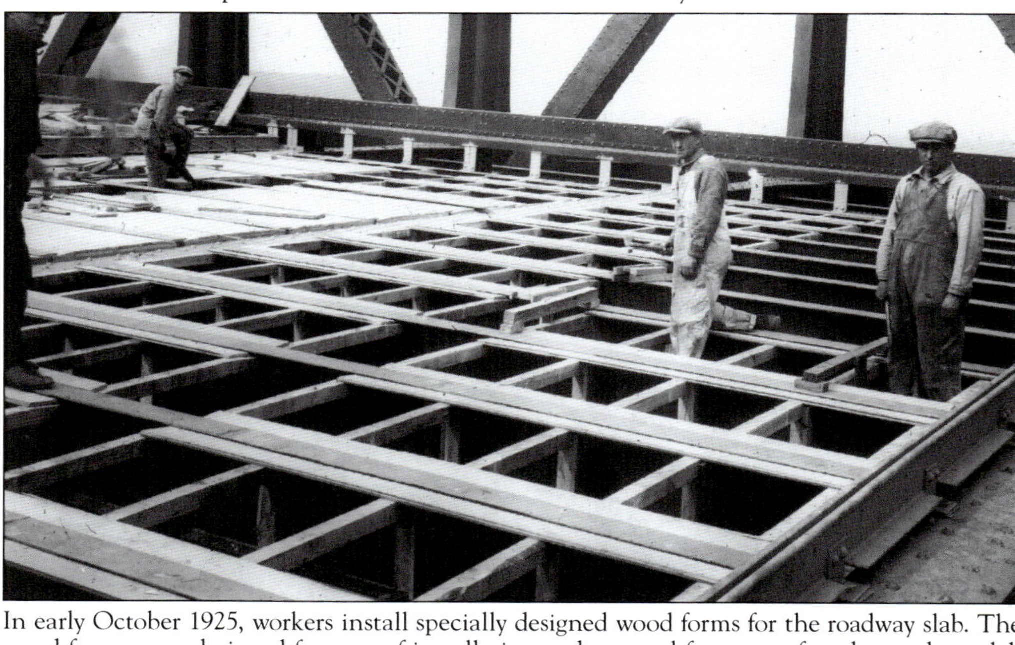

In early October 1925, workers install specially designed wood forms for the roadway slab. The wood forms were designed for ease of installation and removal for reuse after the roadway slab was poured.

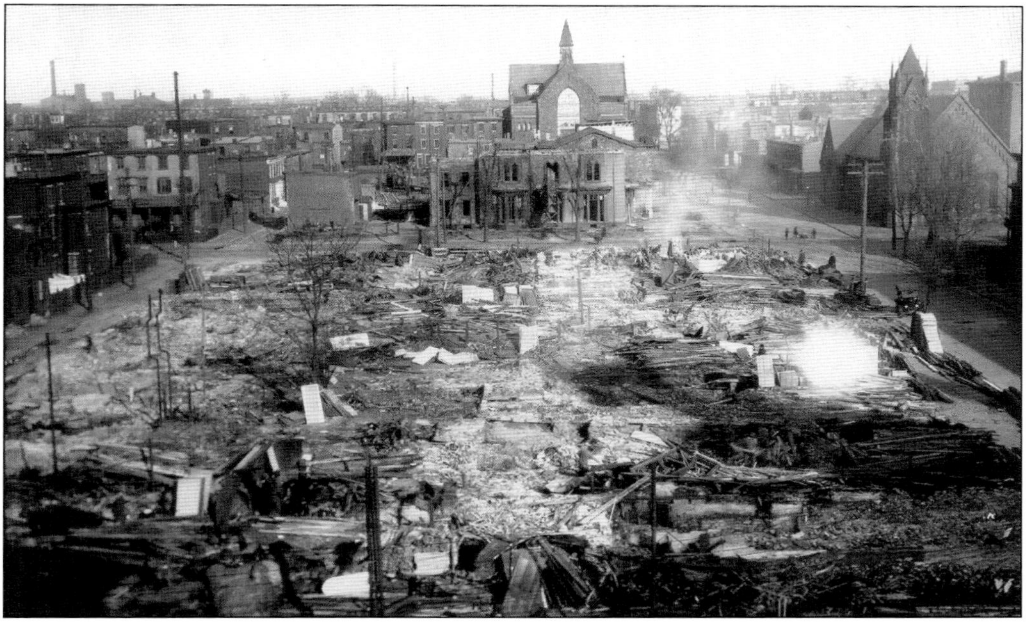

The bridge approach in Camden runs from Delaware Avenue to Third Street between Birch and Pearl Streets, where it turns toward the southeast and terminates at a plaza at Broadway at Penn Street. To construct the approach on this alignment, existing commercial and residential structures had to be razed. In the image above, demolition has progressed to Second Street. Crews have also started demolition work inside the structures in the background, between Second and Third Streets. In the image below, looking east from Second Street, workers are installing sections for Camden's eight girder approaches. Cross-bracing is used to tie the two columns, called bents, together to act as a single unit. According to Charles Carswell, "The supporting steel bents are of a design to harmonize with the main towers of the bridge." In the background, cranes on temporary tracks lift the steel roadway floor beam sections into position. (Below, courtesy of Modjeski and Masters, Inc., Mechanicsburg, PA.)

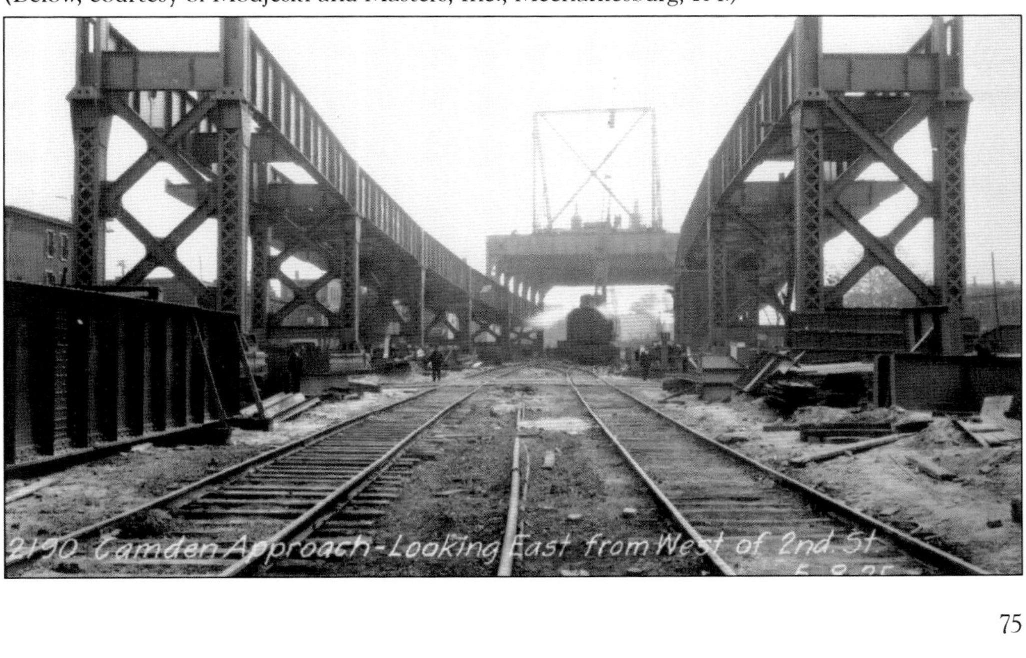

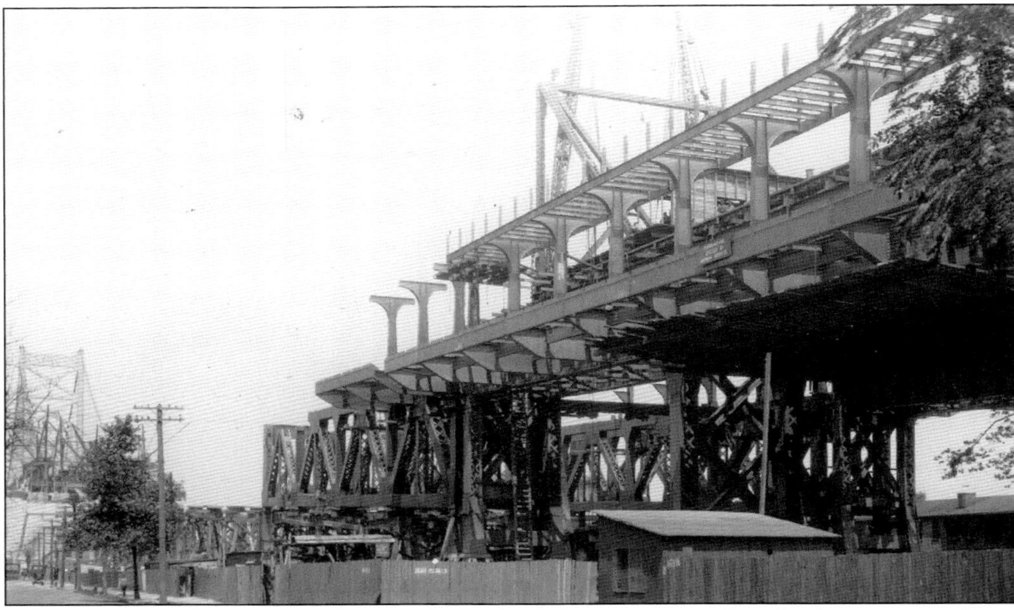

Two months after the previous image, the steel erection process in Camden has progressed west of Second Street. Between Second Street and the anchorage, seven truss spans make up the approach. The truss spans were used in consideration of the roadway height and the spacing between bents.

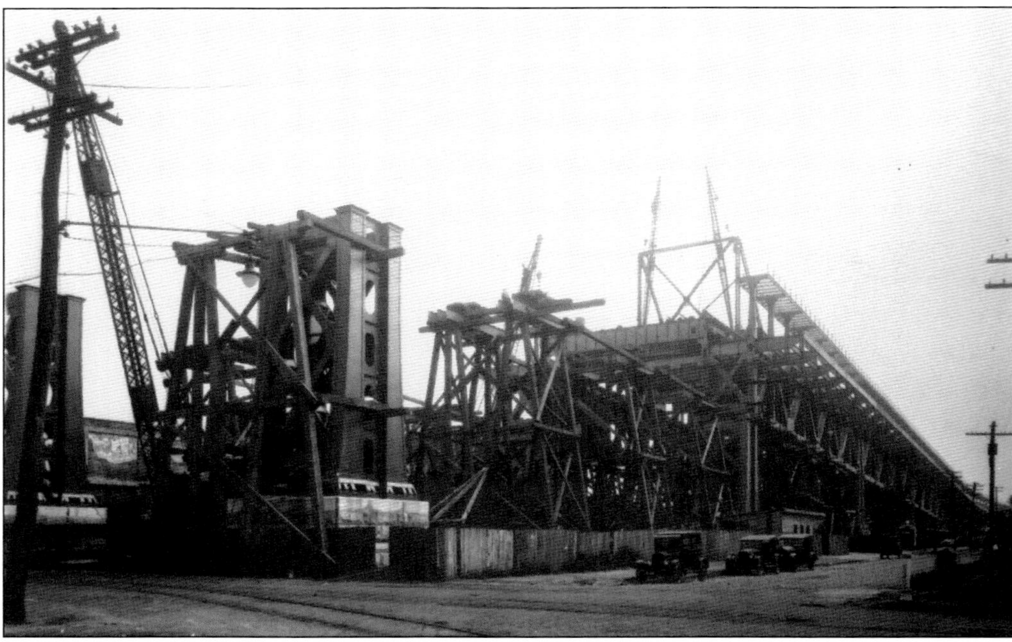

Looking northeast from Delaware Avenue at Pearl Street, work on the Camden approach is nearing Delaware Avenue. As Charles Carswell noted, the bents are designed to harmonize with the main towers, and "the truss spans are similar to those in the suspended span." Each of the steel bents that support the bridge approaches are secured to a foundation of concrete with a granite face, similar to the bridge towers.

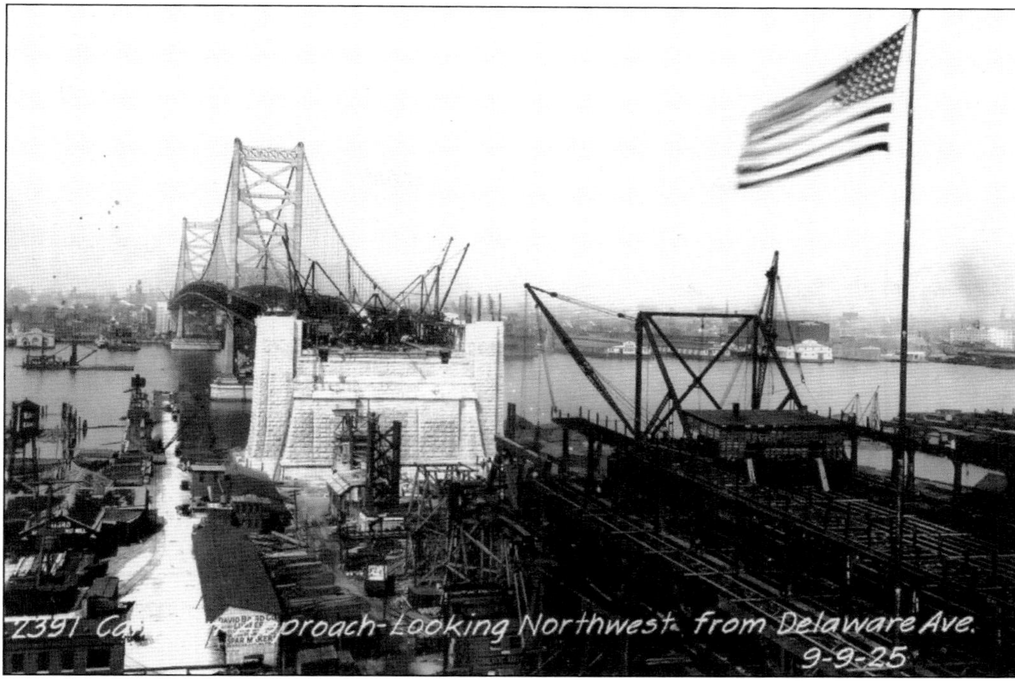

An American flag (with only 48 stars) flies prominently in this September 1925 view of bridge construction. Work crews are completing the anchorage towers, and the truss spans on the Camden approach have progressed across Delaware Avenue. This image was taken from the Victor Talking Machine Company building on Delaware Avenue at Pearl Street.

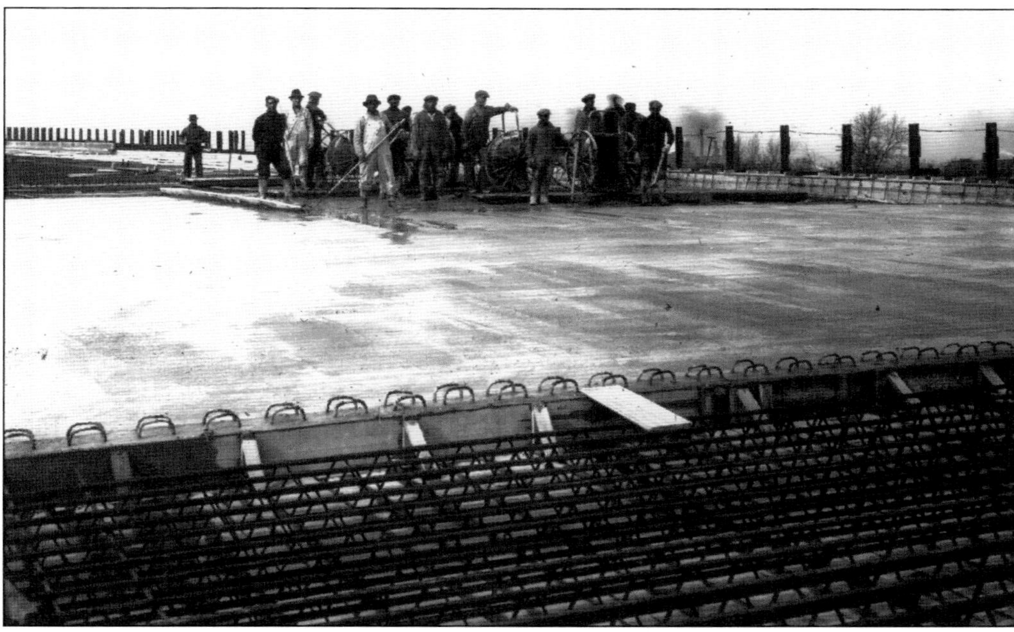

At Third Street in Camden, to strengthen the roadway, workers have installed a reinforcing steel mesh that consists of prefabricated trusses. In this scene from November 1925, concrete is being poured for the roadway deck.

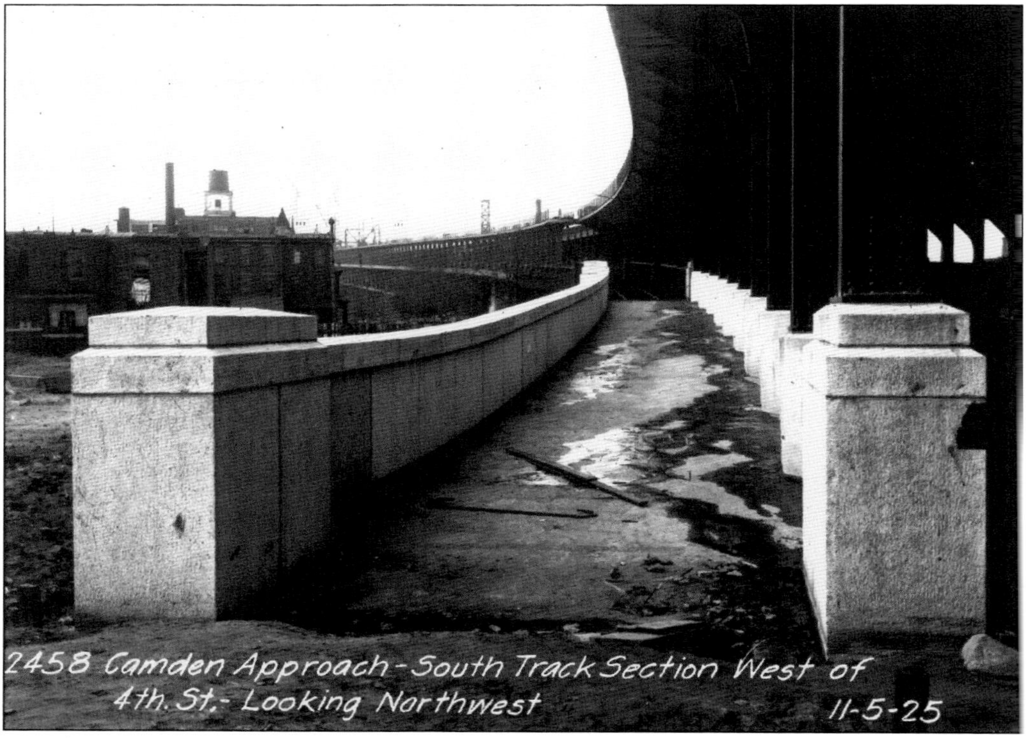

Seen here is the area where the rapid transit's surface line would have descended off the bridge and entered Camden's local streets. Although both pairs of rapid-transit lines descended off the bridge east of the anchorage, only the inner surface-line track areas were constructed to grade. The steel supporting the high-speed rapid-transit tracks ended at Third Street, where the line was planned to connect with an elevated structure in Camden.

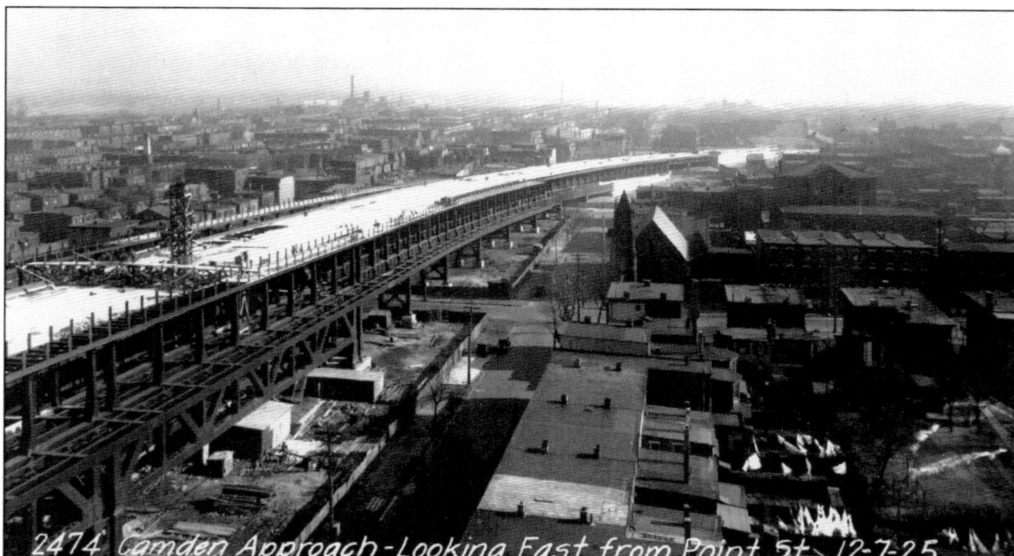

In this view looking east of the Camden approach from the Victor Talking Machine Company building, work on the roadway construction continues as the concrete base is poured.

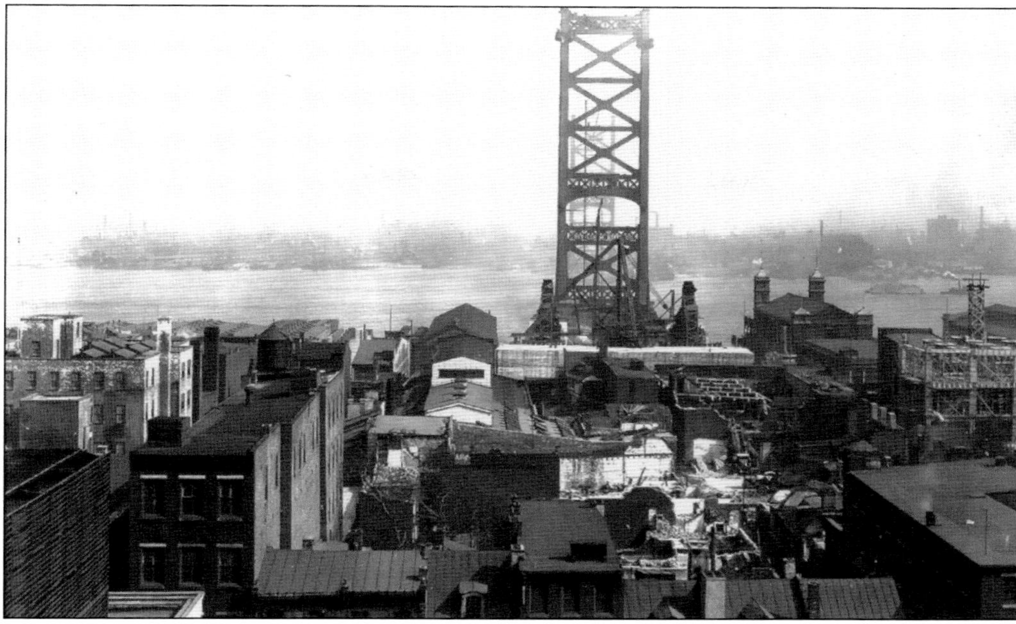

The Philadelphia approach was constructed between Front and Sixth Streets; dozens of affected commercial and residential properties were razed. Seen here, partial demolition is underway between Front and Third Streets. According to Walter S. Andariese, one Philadelphia politician suggested that a circular ramp should be constructed west of the anchorage, so that only one city block would need to be demolished and only the top deck would need snow plowing, because the lower decks would be covered. (Courtesy of Modjeski and Masters, Inc., Mechanicsburg, PA.)

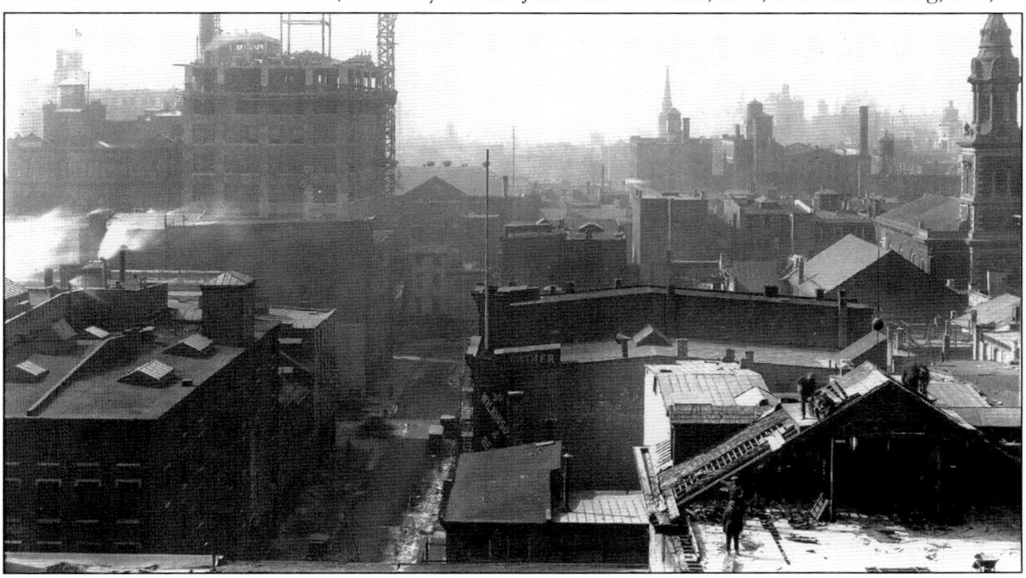

Workers are demolishing the roof of a building on Philadelphia's Second Street. The steeple on the right side of the image is that of St. Augustine's Church. At the bottom and to the left of St. Augustine is the roof of the historic St. George's Church. Original plans called for St. George Church's demolition, creating an uproar that brought a court battle and led engineers to shift the approach slightly to the south, sparing the church.

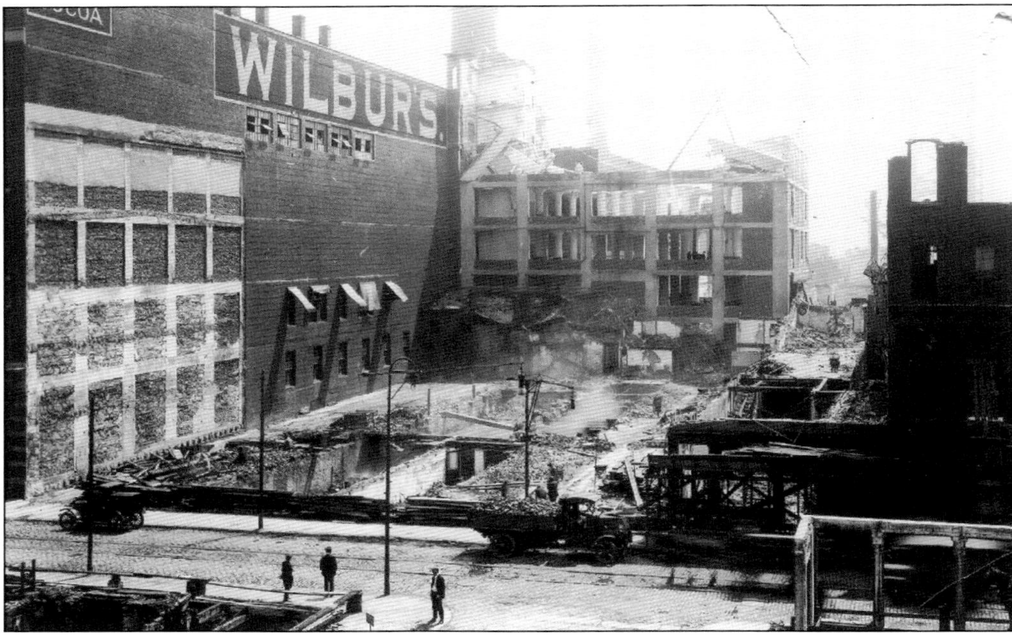

Wilbur's Chocolate, between Second and Third Streets in Philadelphia, is seen here. In 1924, a portion of the manufacturing plant was demolished to make room for the bridge approach. The company continued to operate at this site after the bridge opened. In 1930, the company moved its manufacturing operations to Lititz, Pennsylvania. The building is now a condominium called the "Chocolate Works." Although faded considerably, the Wilbur's sign can still be seen today on the side of the building.

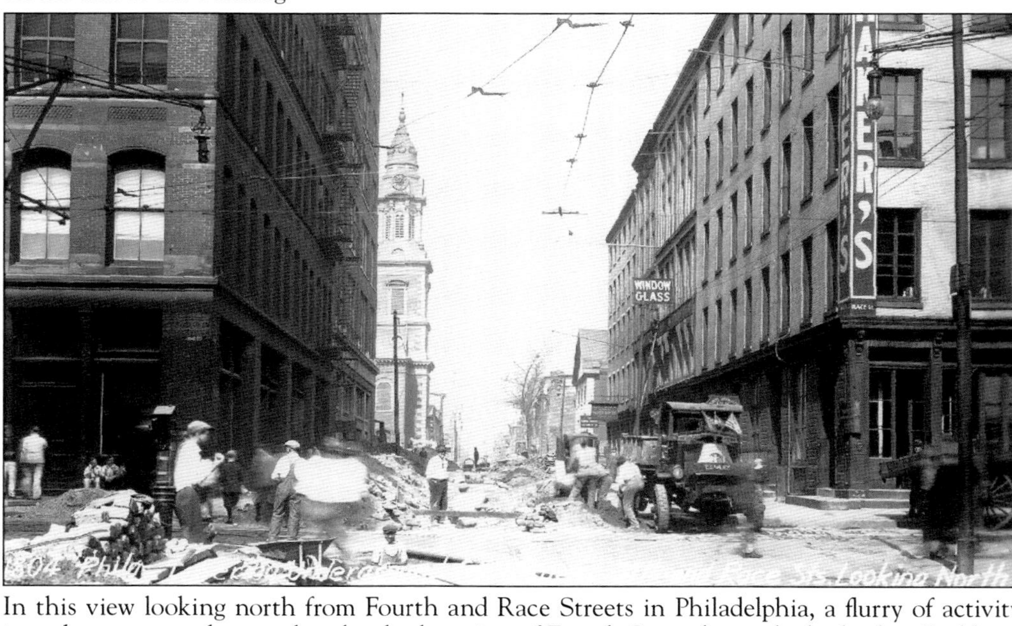

In this view looking north from Fourth and Race Streets in Philadelphia, a flurry of activity is underway as workers undertake the lowering of Fourth Street beneath the bridge. Buildings along Fourth Street between Vine and Race Streets had entrances modified to accommodate the new street-level accesses. In the background, the steeple of St. Augustine's Church can be seen.

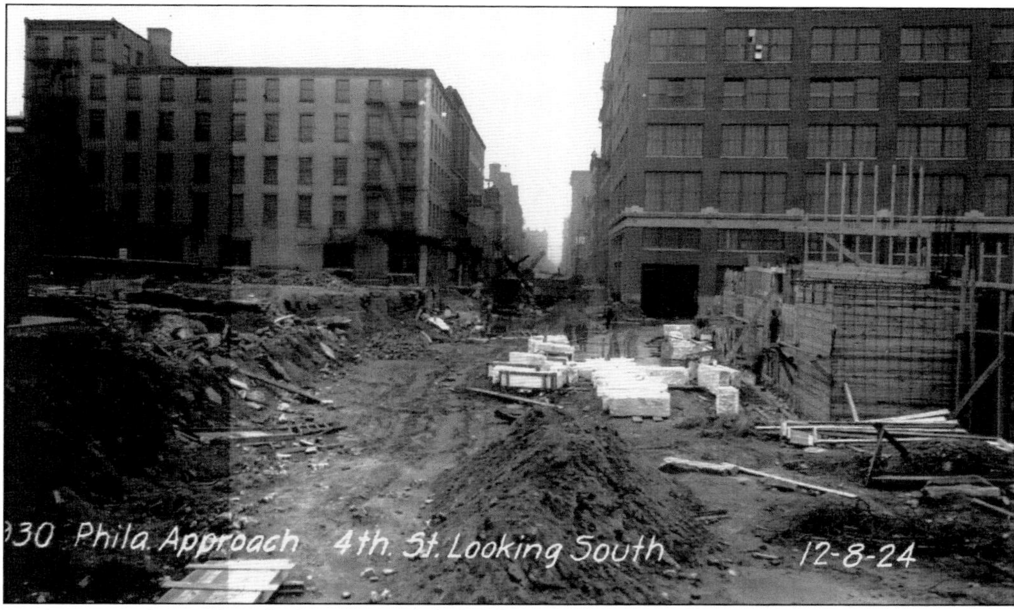

Looking south on Fourth Street in Philadelphia, work progresses on a structure beneath the bridge that will serve as a maintenance building. With the lowering of Fourth Street, the front entrance to St. Augustine's Church (right, outside of picture) was also lowered to match the new elevation. Across the street (extreme left), St. George's Church retained the same entrance; however, a new retaining wall was constructed around the basement. A stairway was constructed along Fourth Street to the front entrance.

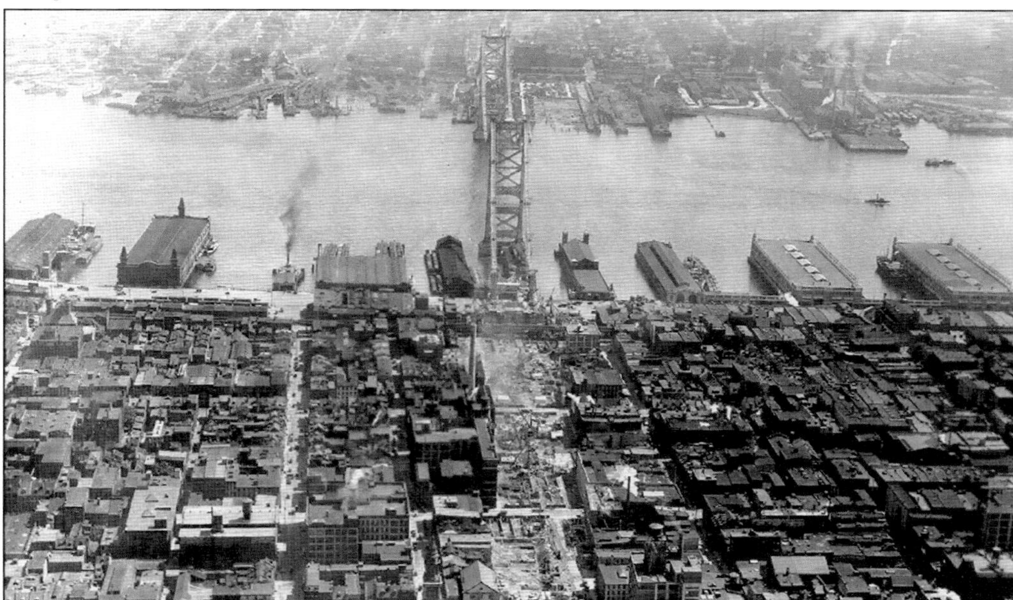

This aerial photograph of Philadelphia, showing the demolition work between Sixth and Front Streets, provides insight as to the precision used to thread the bridge approach through the neighborhood. The DRBJC wanted to minimize the impact on the neighborhood and demolished buildings only as required.

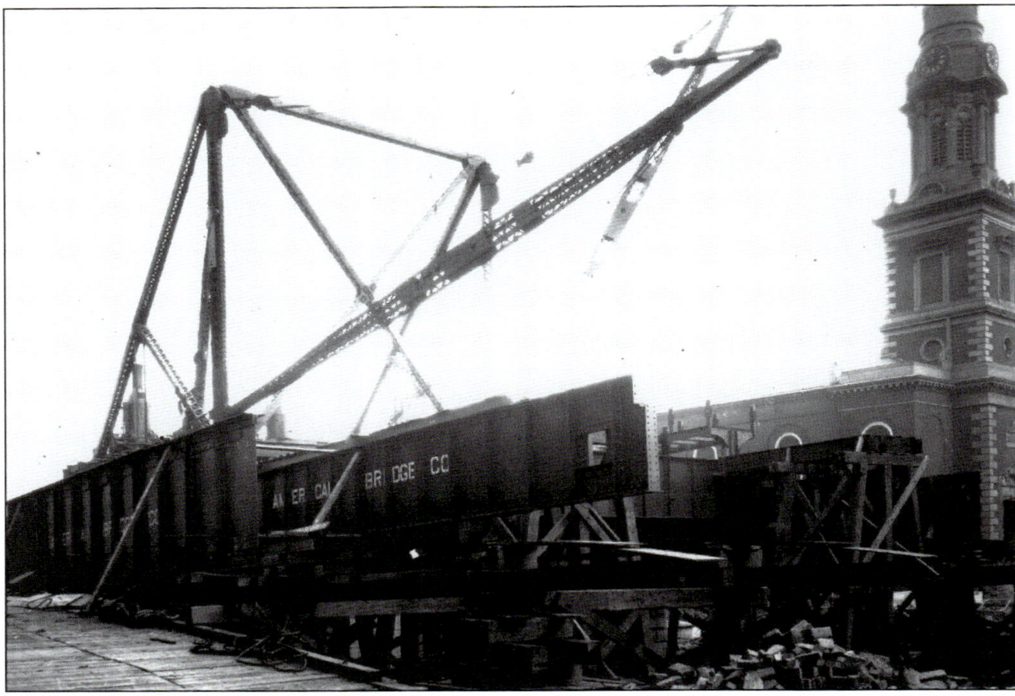

In May 1925, workers started to erect the structural steel of the Philadelphia approach. In this photograph, cranes are moving floor beam sections into position just east of Fourth Street.

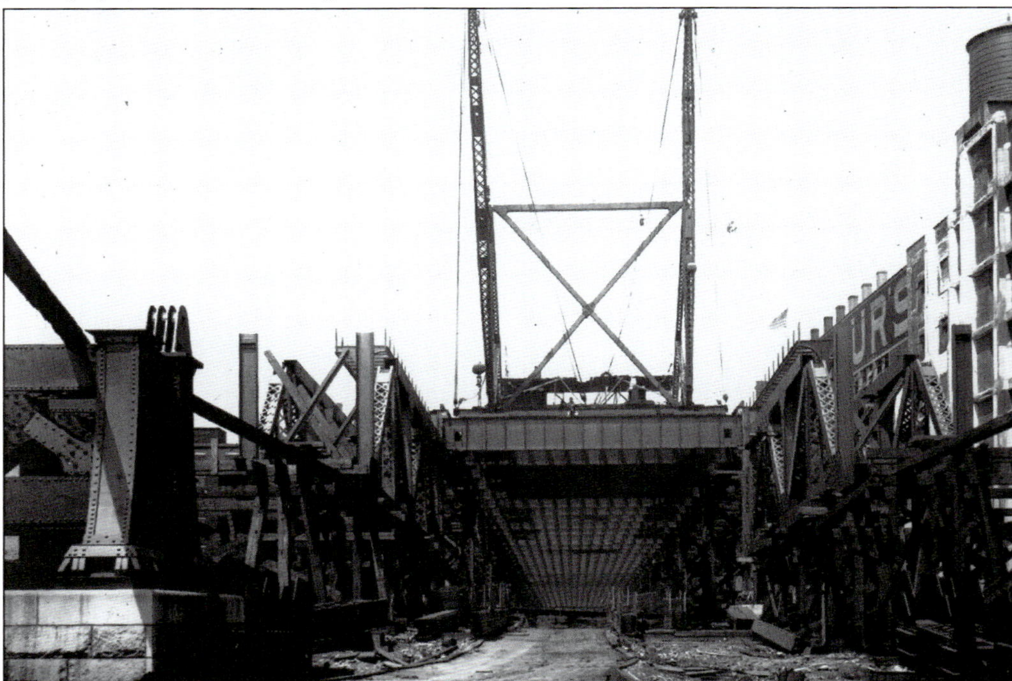

In this view from July 1925, construction of the approach has progressed past Third Street in Philadelphia, where it transitions from the girder to the truss spans.

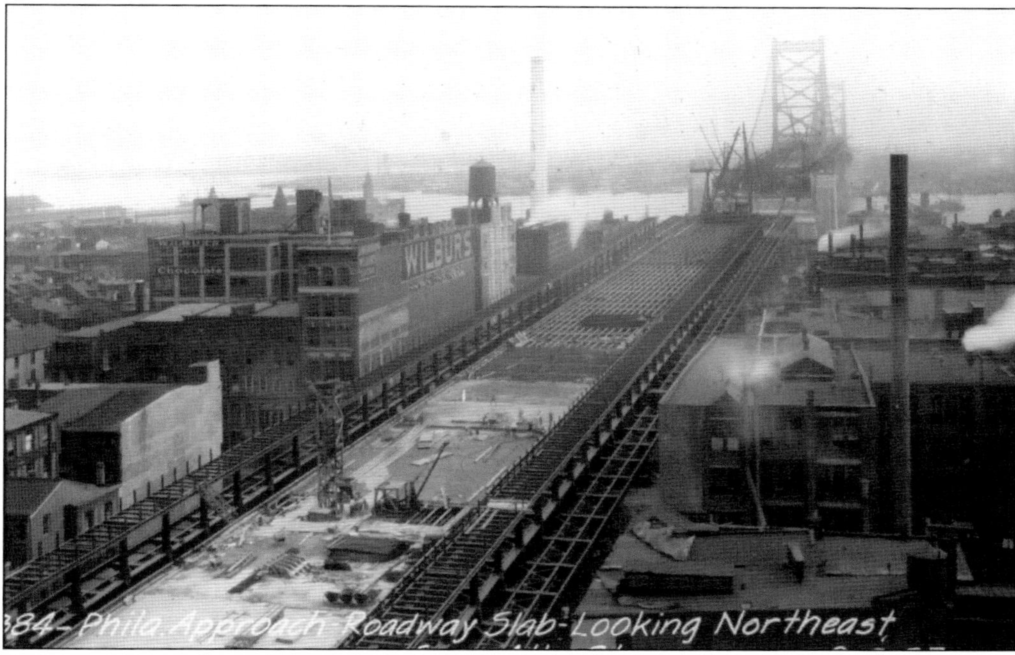

As bridge approach steel erection continues at Front Street in Philadelphia, reinforcing steel trusses for the roadway's concrete deck have been installed across Third Street. Behind the workers installing the roadway reinforcement, others are pouring concrete for the roadway slab.

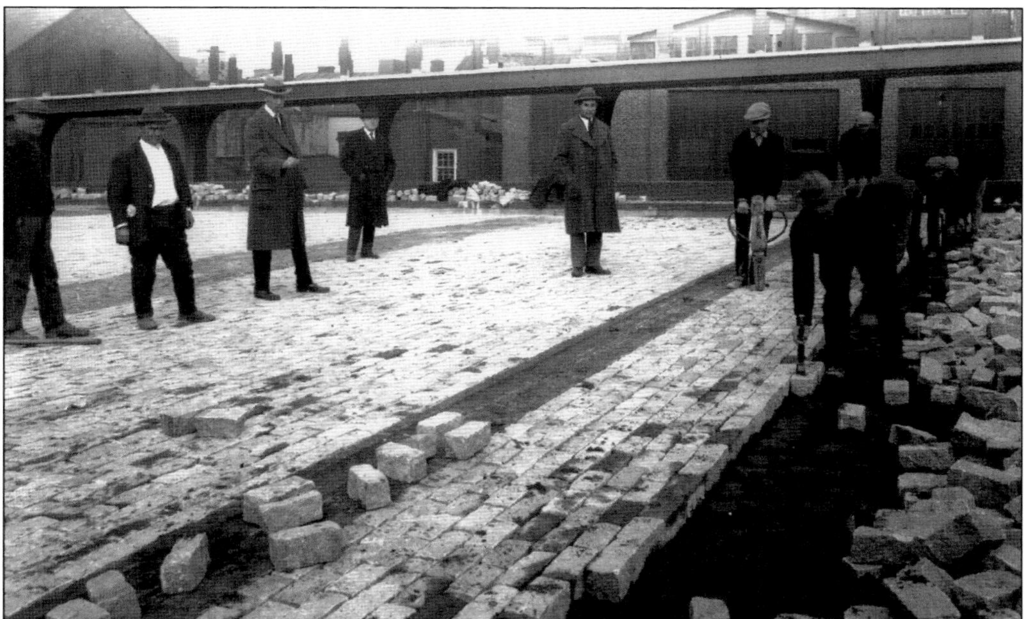

Along the Camden and Philadelphia approaches and across the anchorages, granite block pavers are installed on the roadway deck to assist with waterproofing. After the granite blocks were installed, an asphalt overlay was applied to fill in the voids between the blocks. A second asphalt layer was applied for the roadway surface.

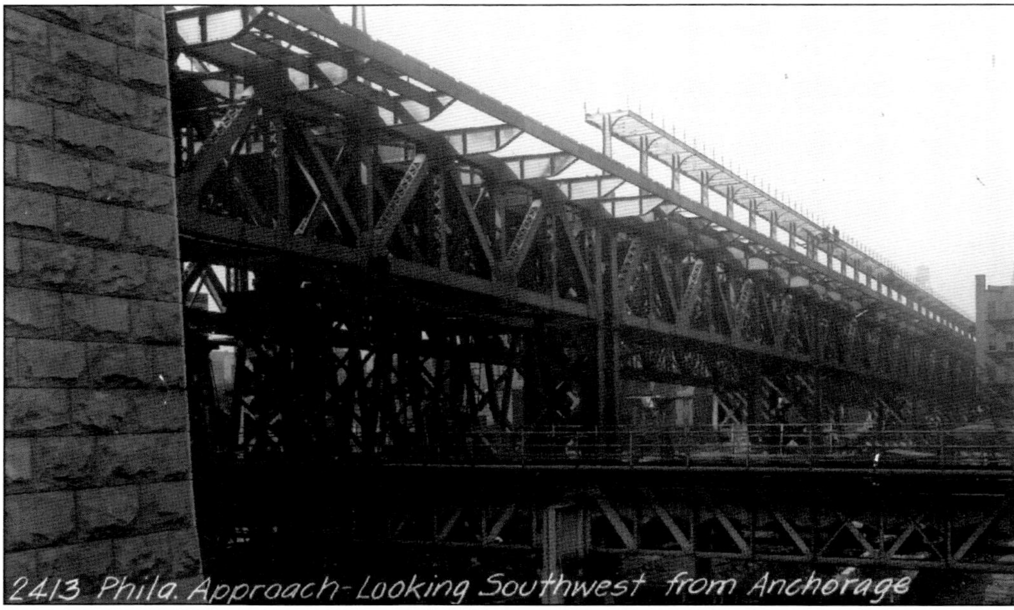

Looking west of the Philadelphia anchorage, structural steel erection for the approach has reached the anchorage. In this image, one can get a sense of the proximity of the anchorage, the approach steel, and the Frankford Elevated track structure. If taken today, a photograph of this location would also include six lanes of Interstate 95, ramps to the Vine Street Expressway, and a reconstructed Market-Frankford Elevated.

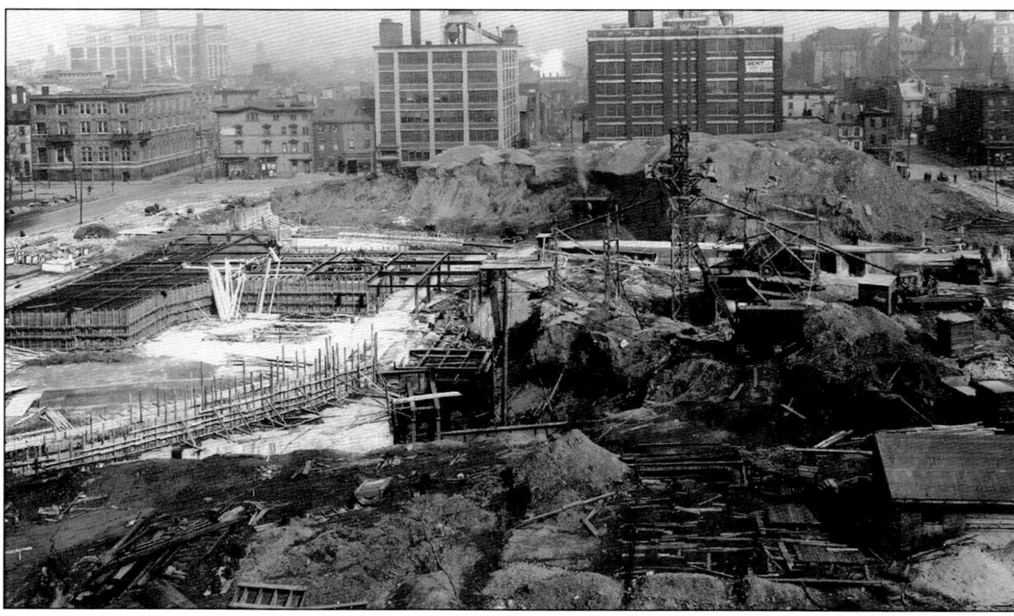

Between Fifth and Sixth Streets, the approach to the bridge loops around the Philadelphia plaza, a two-level structure that was to serve as a rapid-transit station. Tracks descending off the bridge would have entered the plaza on the lower level, which was planned to be a trolley station. The tracks would have been configured to allow the trolleys to loop around and return to New Jersey.

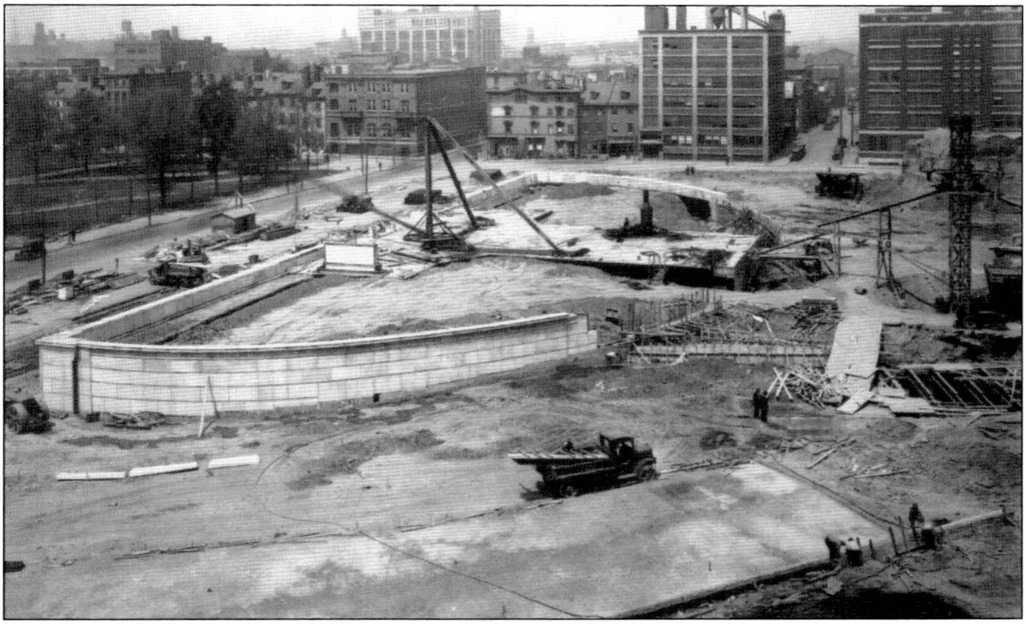

Construction on the Philadelphia plaza is nearing completion. The west portion is complete, whereas the east wall and steps still require work. Crews have begun work on the roadway leading from Race Street to the bridge. Concrete still must be poured for the sidewalk surrounding the plaza and the tunnel between the plaza and the bridge (middle right). (Courtesy of Modjeski and Masters, Inc., Mechanicsburg, PA.)

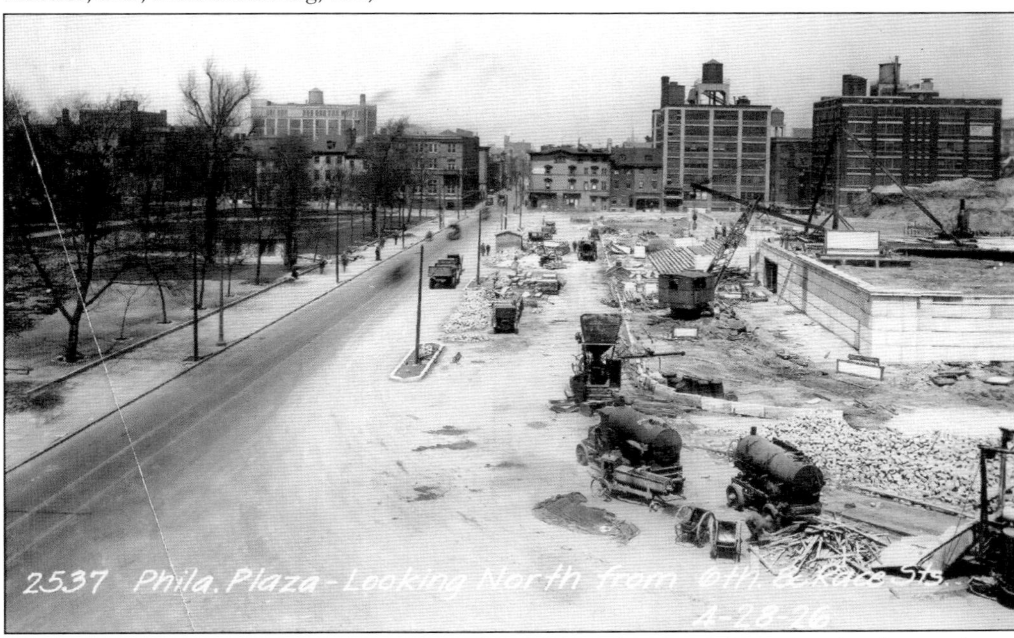

The tremendous amount of activity at the Philadelphia plaza is made evident by the number of construction vehicles on Sixth Street. Between Race and Vine Streets, Sixth Street was doubled in width to accommodate the traffic leading to and coming off the bridge. An early plan for the plaza called for construction of a tunnel under Sixth Street between Vine and Race Streets.

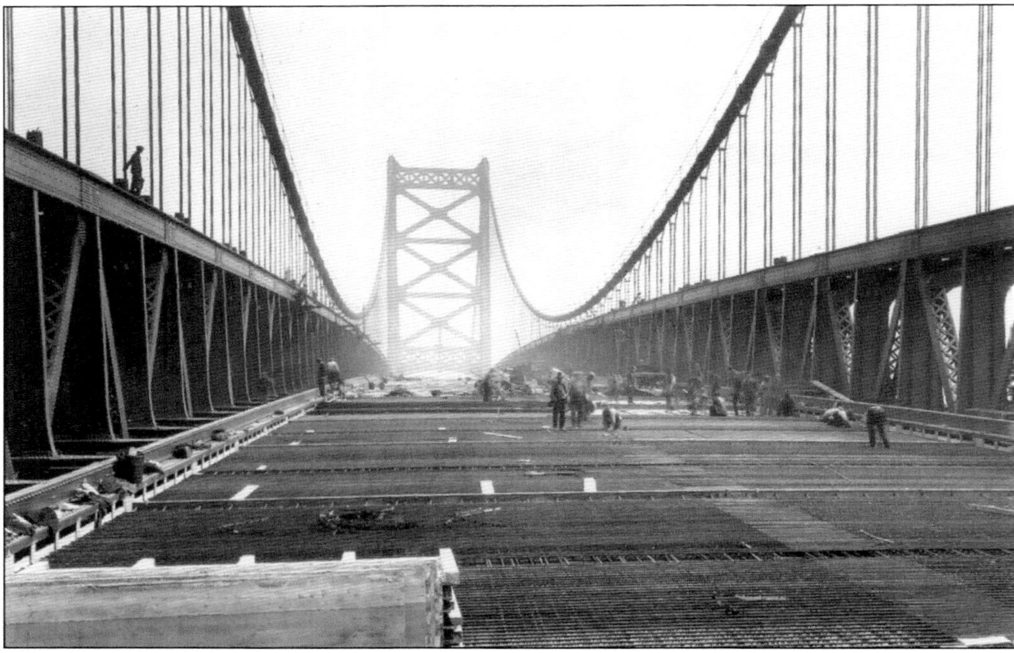

After the wood forms were installed, workers placed a mesh of reinforcing steel on the main span. The reinforcing steel consists of small trusses that were fabricated off-site. The reinforcement was used to strengthen the concrete that formed the base of the roadway.

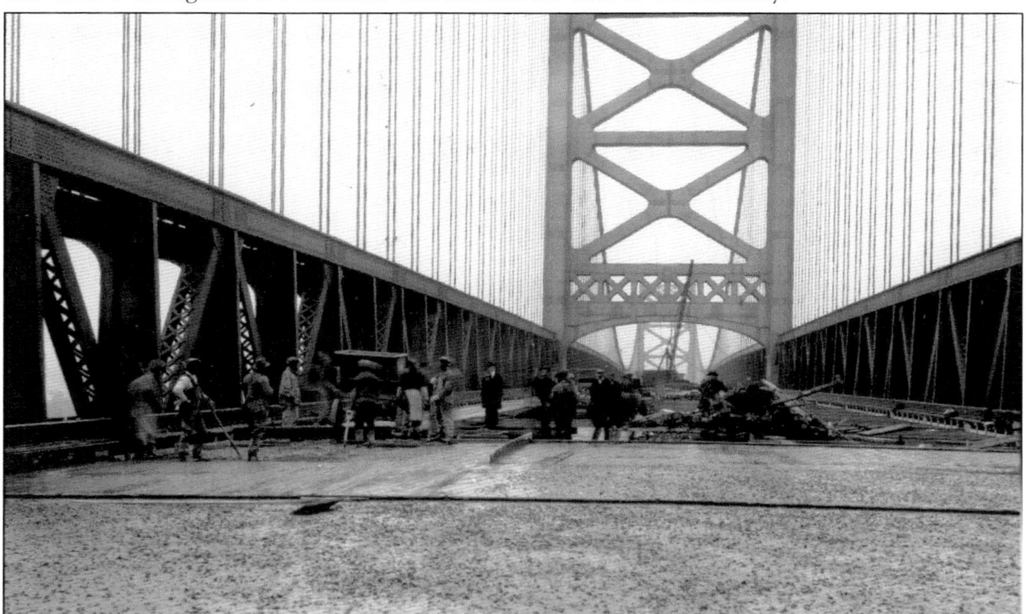

Pouring of the concrete base for the bridge roadway is underway in this image. The concrete was mixed in plants beneath the bridge and hoisted up to the roadway, then transported to the appropriate roadway section. Workers are milling around the truck used to haul the concrete along the Camden side span. The truck operated across the track used by the traveler crane during roadway steel erection; as work progressed, the temporary tracks were removed.

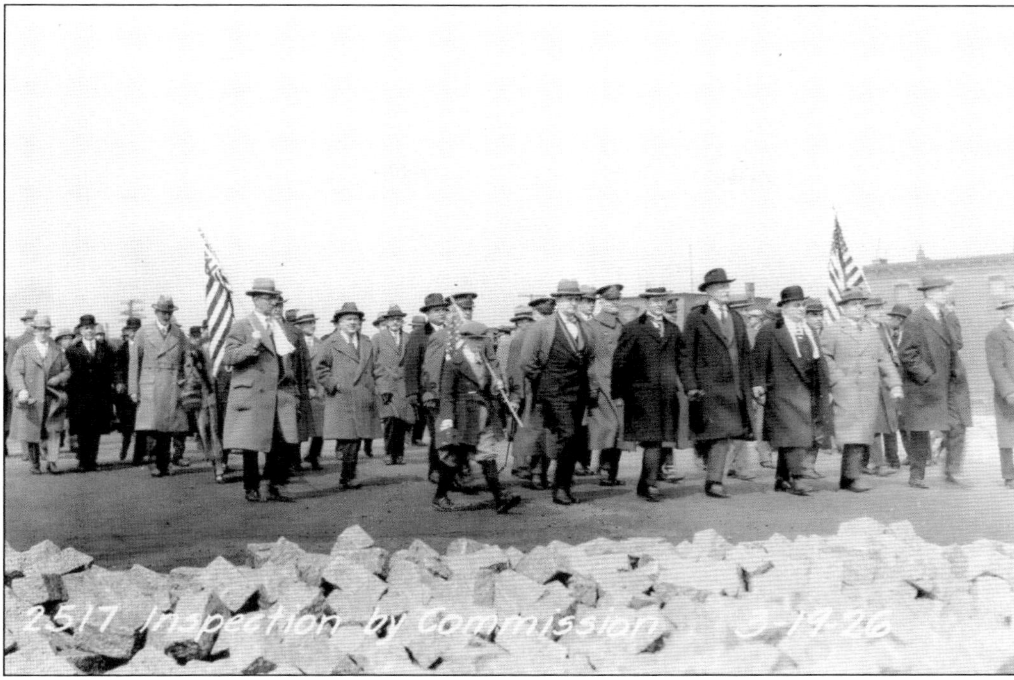

On March 19, 1926, members of the DRBJC and other officials arrived at the bridge for an inspection. This site visit followed the completion of concrete pouring for the roadway surface. As seen in the picture below, the inspection party included Gov. A. Harry Moore of New Jersey (third from left, front row) and Gov. Gifford Pinchot of Pennsylvania (fourth from left, front row).

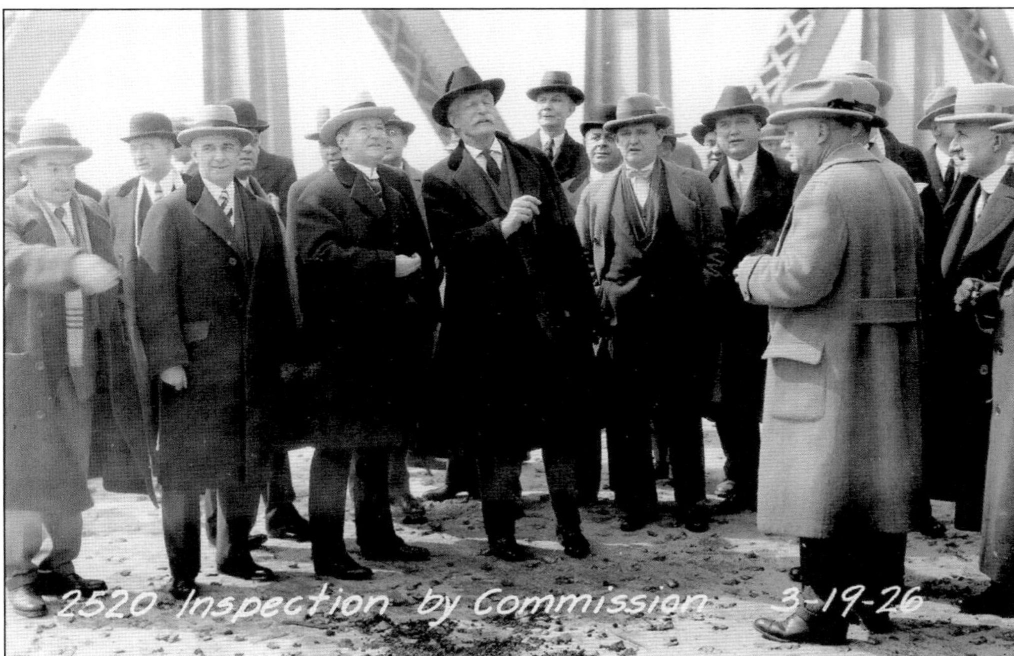

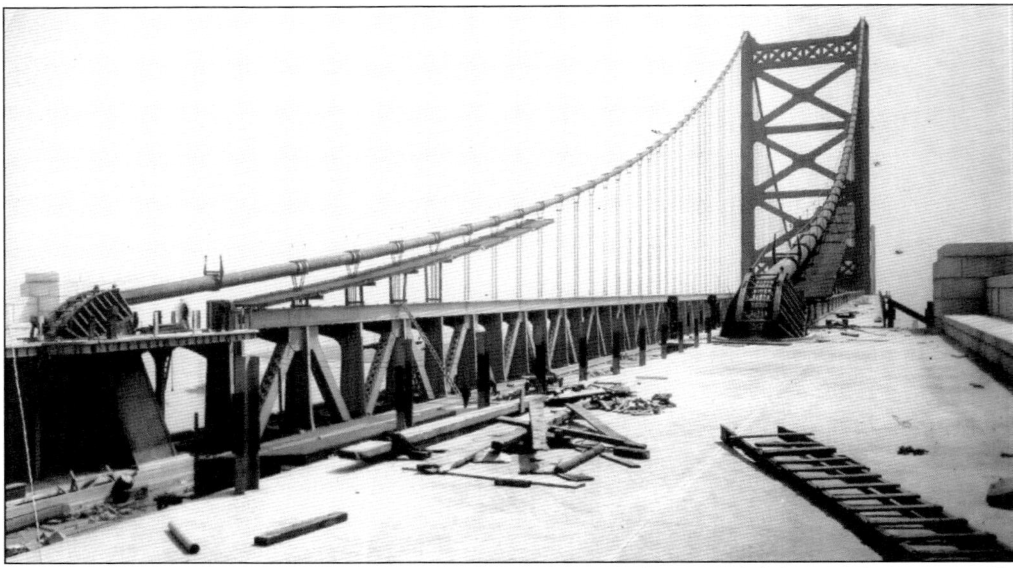

Above, construction debris litters the walkway area at the Philadelphia anchorage. In the background, working platforms are being installed beneath the main cables for use during the cable-wrapping process, which is seen in the image below. According to Charles Carswell, "The cable wrapping contract provided for the important work of encasing the compacted cables within an armor of steel wire." This process involved the wrapping of a single wire over the exposed surface, which began at the low points of the cable then moved upward toward the towers. This method ensured that the wires remained in continuous contact with the adjacent wrapping wire. After the cables were wrapped, the exterior wire was painted to provide another layer of protection from the elements.

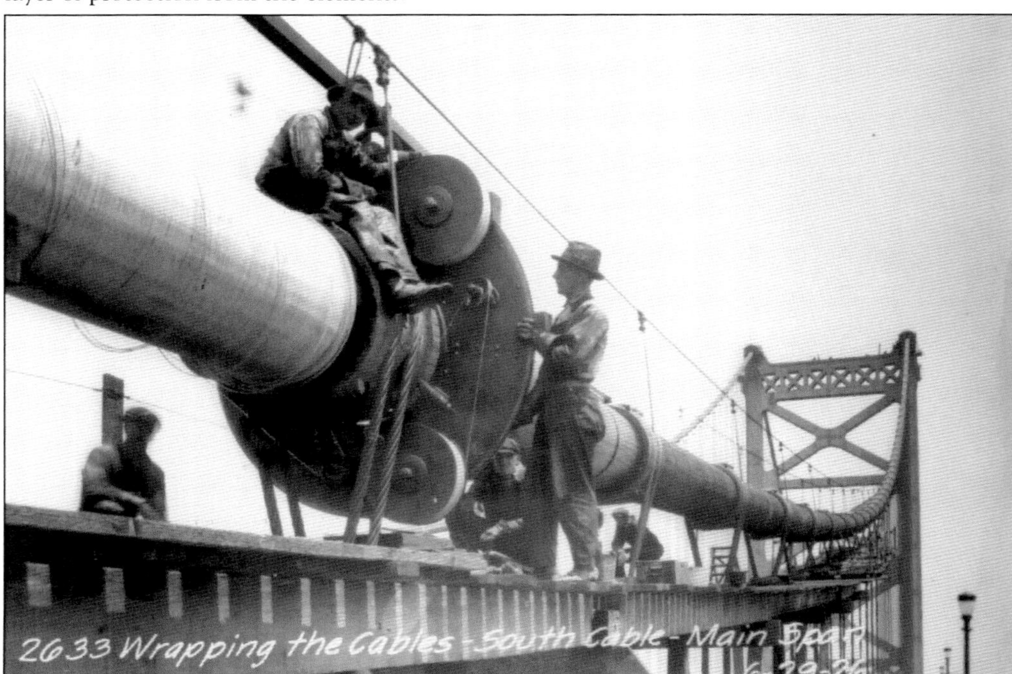

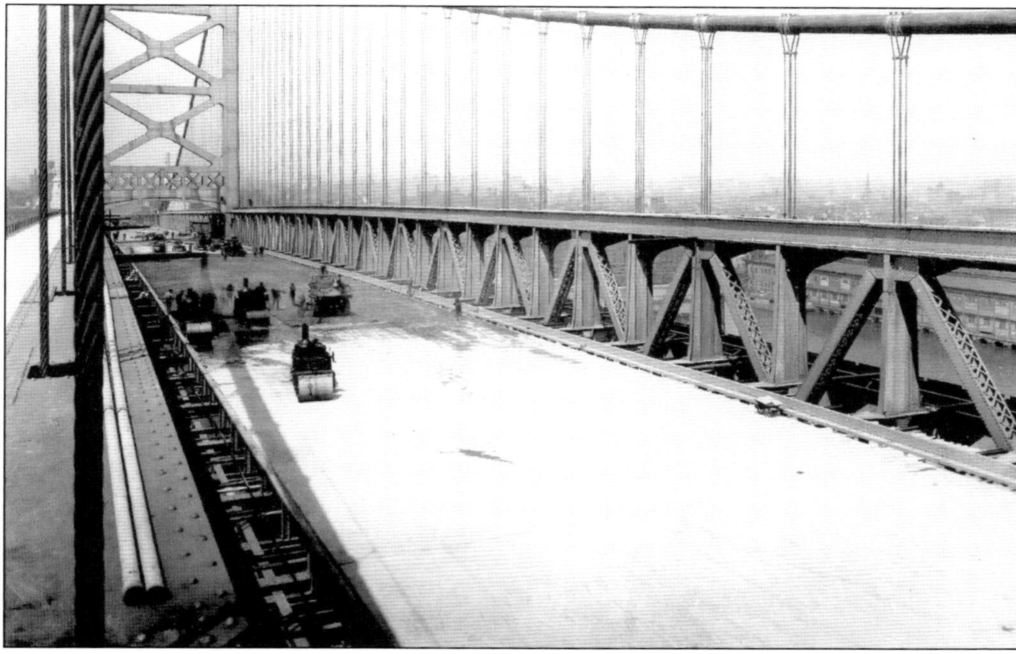

With just over two months remaining until the bridge was scheduled to open to the public, work remains to be completed. This image shows workers applying a 2.5-inch layer of asphalt on top of the roadway's concrete base on the main span.

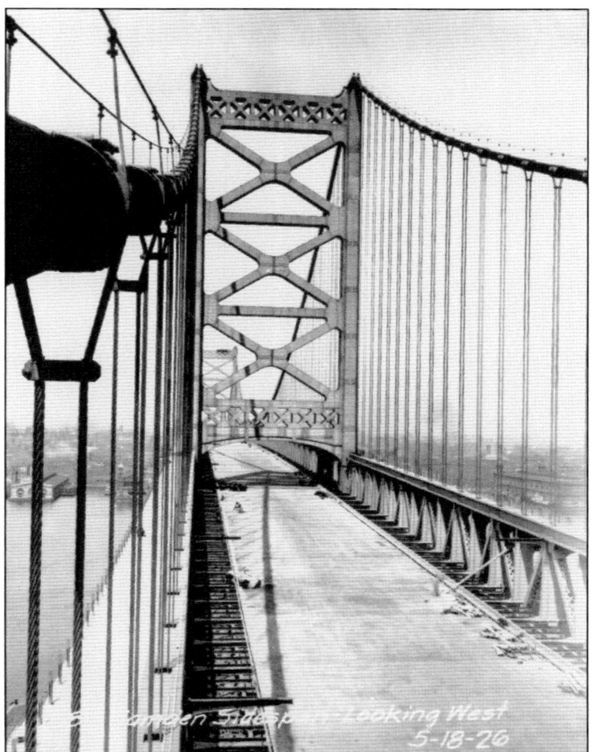

In this image, the bridge opening is six weeks away, and it is almost ready. Remaining work includes grating for the rapid transit's surface track area (adjacent to the roadway), lighting, wrapping the cable, and the pedestrian walkway's handrails and panel sections. (Courtesy of Modjeski and Masters, Inc., Mechanicsburg, PA.)

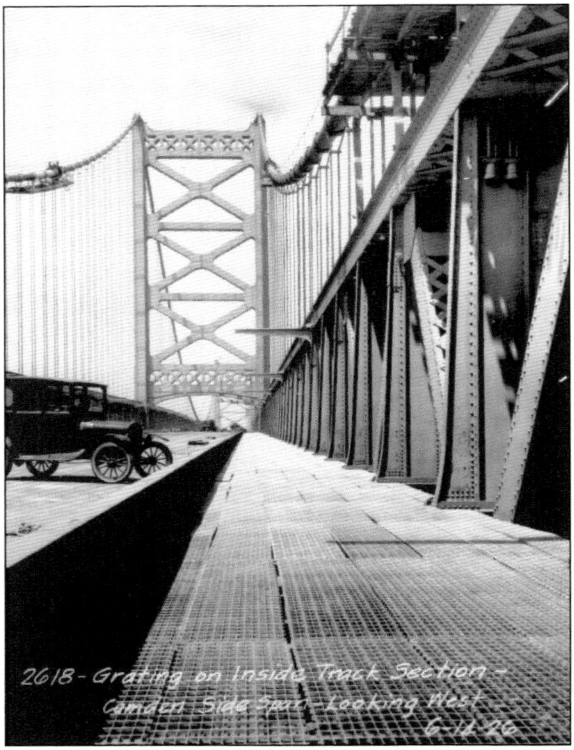

Taken 16 days before the bridge was opened to the public, this image shows the track area for the rapid-transit surface lines. The original intention, never realized, was for trolleys to operate within this area. High-speed rapid-transit tracks were located on the other side of the stiffening truss "wall" (right). The cable-wrapping process was still ongoing (upper left). As the cable wrapping progressed, platforms were moved ahead of the mechanical wrapper. (Courtesy of Modjeski and Masters, Inc., Mechanicsburg, PA.)

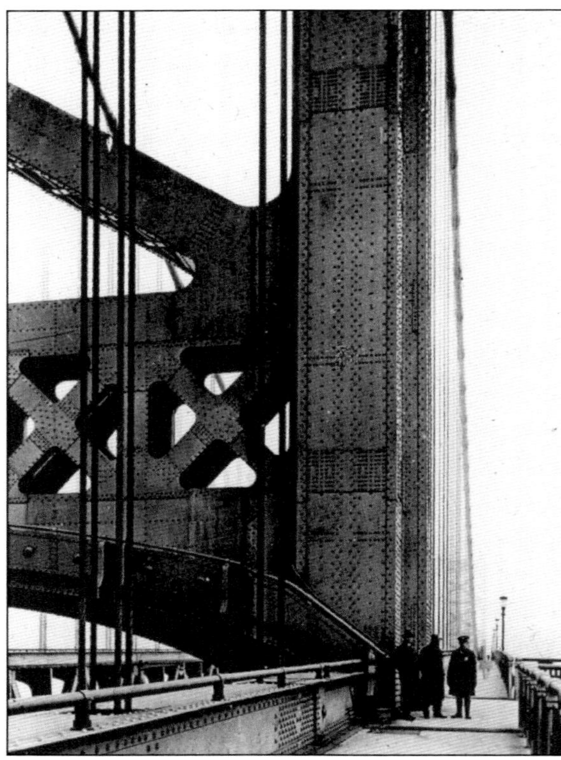

In 1926, work was complete on the walkways, as well as the crossover along the tower face. The size of the tower's roadway portal strut and transverse bracing are evident when compared with the individuals standing on the walkway at the tower.

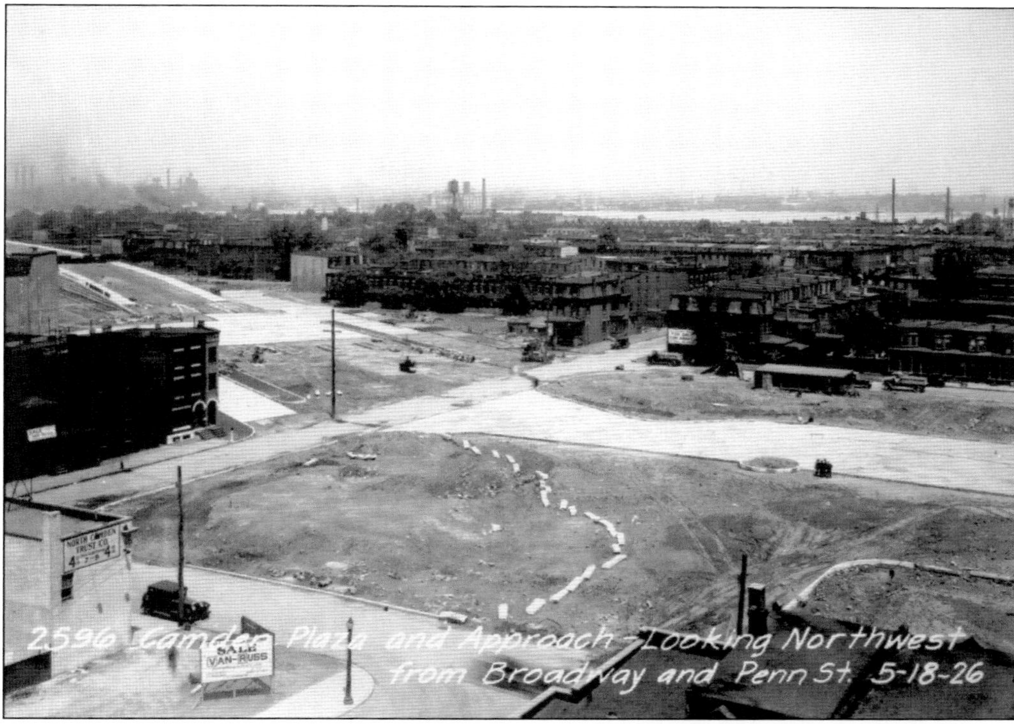

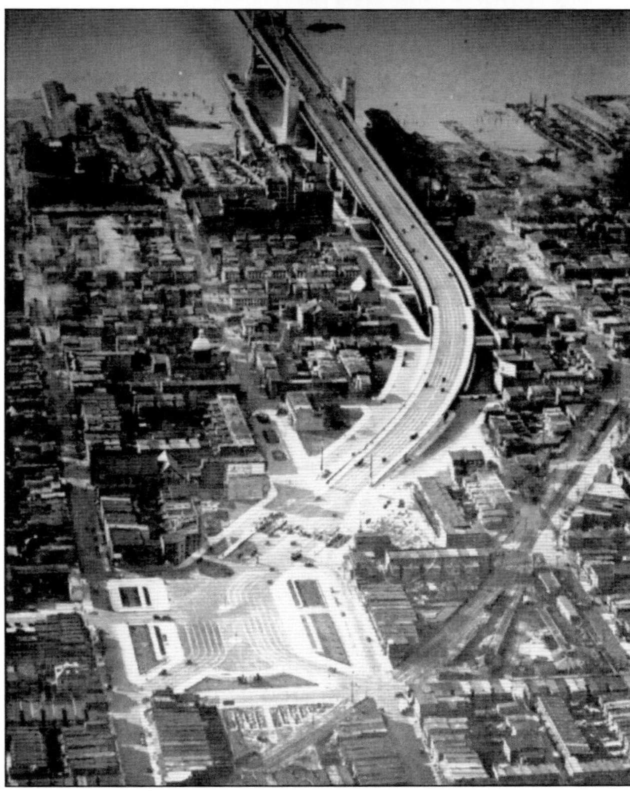

With the bridge's opening only six weeks away, work crews are continuing with construction at the Camden plaza in the image above. The plaza, which took up a full city block, was designed to handle traffic approaching the bridge from Linden Street and Broadway (which had been extended two blocks to connect to the bridge plaza) and traffic coming off the bridge and traveling to Broadway and Penn Street. In the image above (upper left), work on the toll plaza is underway, as concrete islands have been built to serve as a base for the toll booths. After the bridge opened, one can see in the aerial image at right the traffic flow to and from the bridge. Traffic movements were handled by a traffic signal that guided motorists through the intersection. (Above, courtesy of Modjeski and Masters, Inc., Mechanicsburg, PA.)

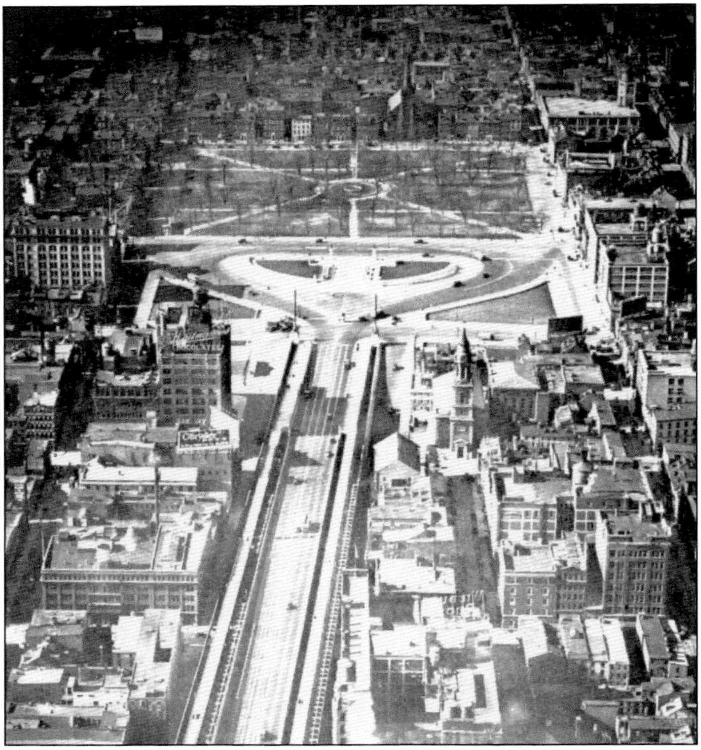

In this aerial view of the approach and Philadelphia plaza taken within a year of the bridge's opening, one can see the variety of options available for traffic to move off the bridge into the city's street system. Not long after this picture was taken, a traffic light was installed at the base of the bridge to allow Fifth Street traffic to move north past bridge traffic, which created considerable tie-ups.

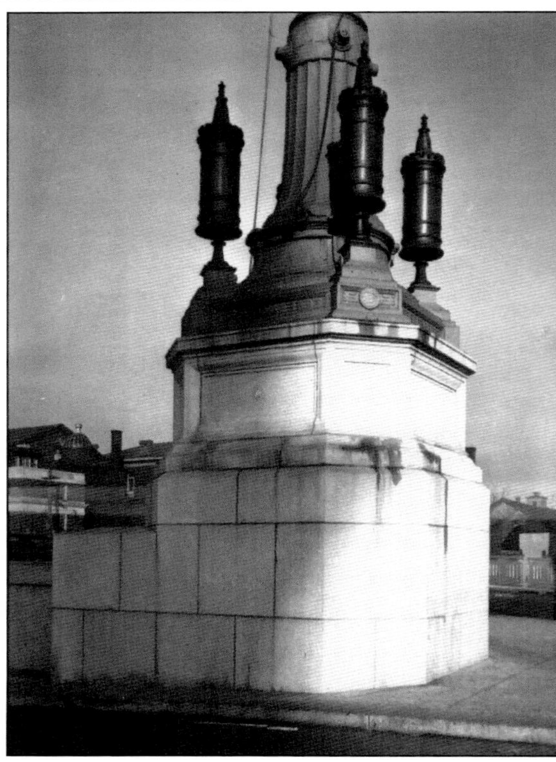

At the base of the bridge in both Camden and Philadelphia, the "Winged Victory" angels stood on 75-foot steel masts that were mounted on cast-iron bases secured to a granite-faced concrete foundation. The angels stood watch over the bridge for almost 20 years until they, as well as the foundations, were removed in 1949, when the bridge's roadway was widened over the two unused rapid-transit surface track areas.

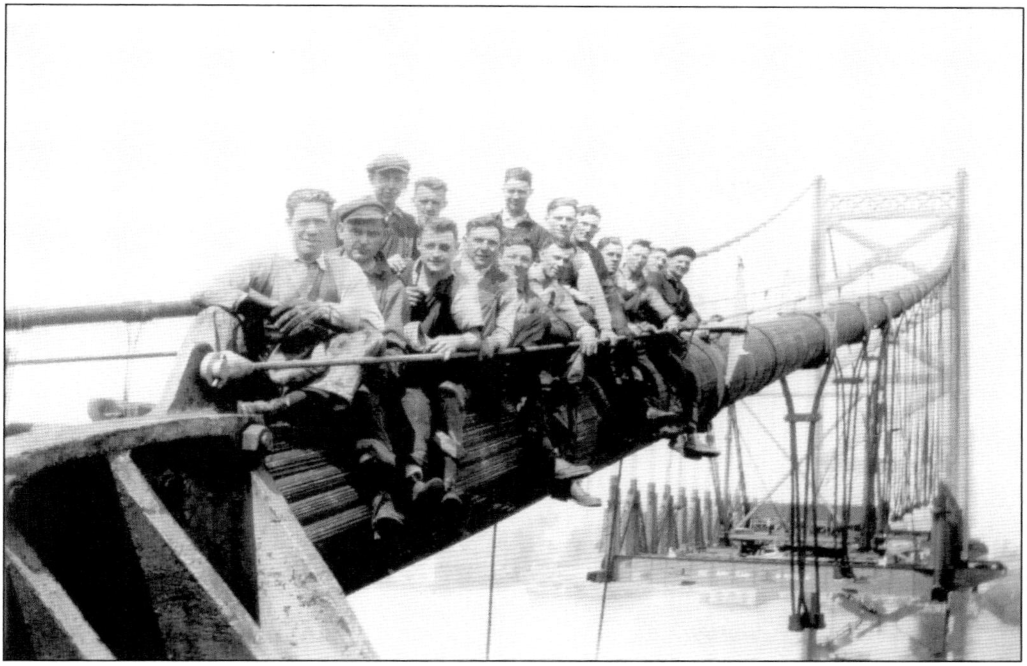

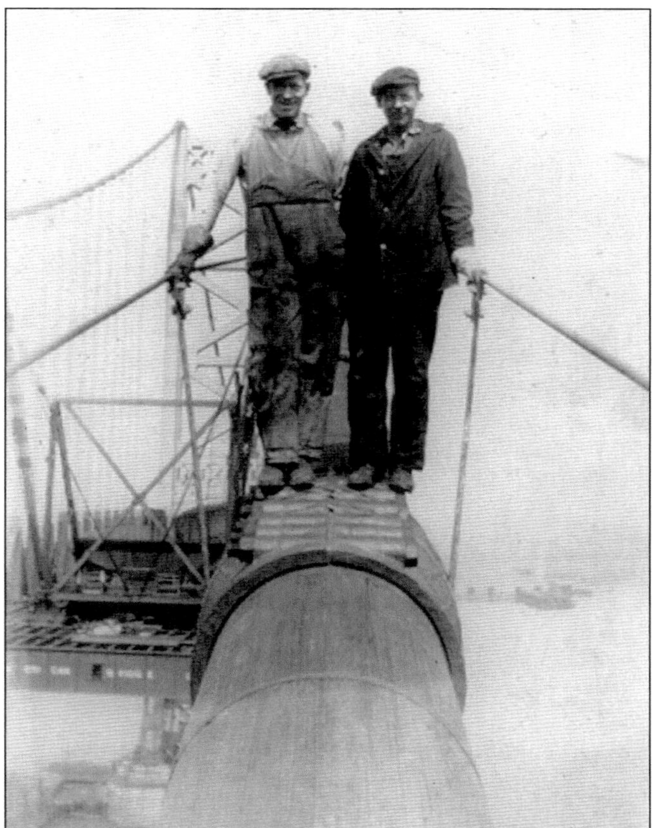

During roadway construction, workers take a break to pose for a picture on the main cable. At the height of construction, 1,300 men worked on the bridge. Some 15 workers died during its construction, mostly by falls from the structure. A suggestion of the rough lives these men may have faced can be seen in the above image, where the third man seated from the left has a black eye. (Courtesy of the estate of Helen Kellmann Meyer, widow of Howard J. Meyer.)

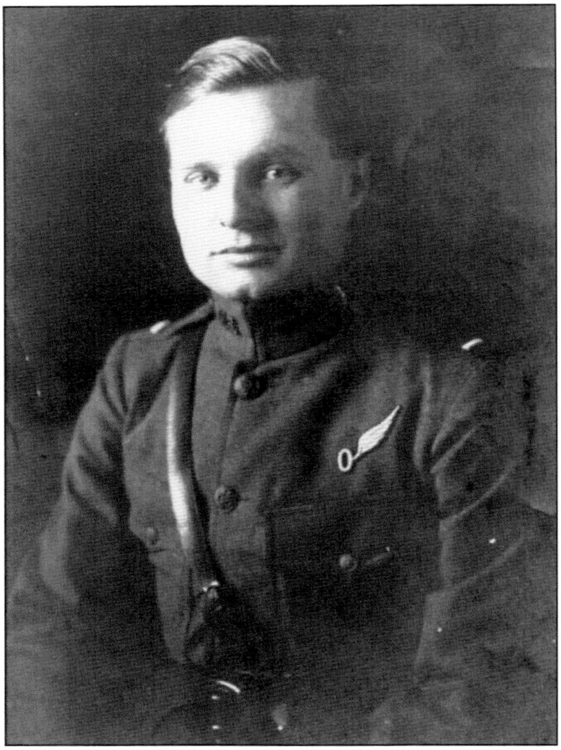

One of the workers who died during construction was Capt. Howard Meyer of Camden. He died in May 1926 after falling from the bridge approach in Camden when an acetylene torch he was using exploded. Captain Meyer was a pilot during World War I, and he was awarded a Distinguished Conduct Medal and the French Croix de Guerre. As part of his funeral service with full military honors, an airplane dropped a wreath on his grave at Arlington Cemetery in Pennsauken. (Courtesy of the estate of Helen Kellmann Meyer, widow of Howard J. Meyer.)

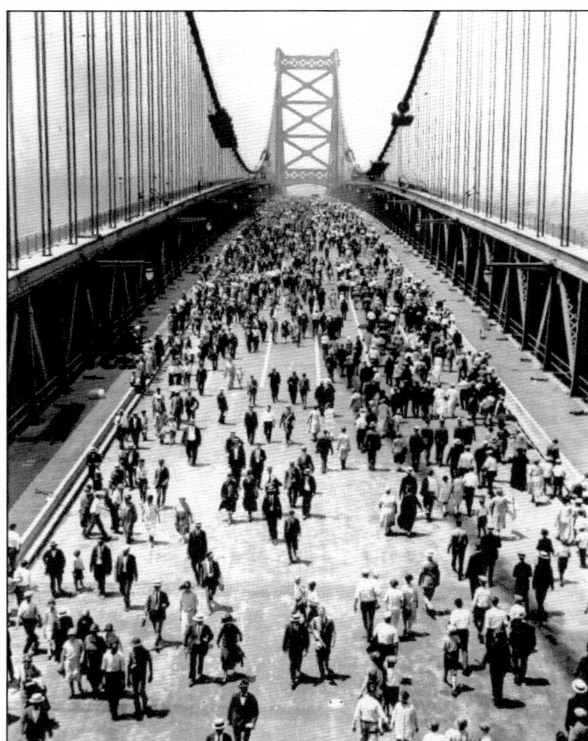

On July 1, 1926, the Delaware River Bridge was opened to pedestrians following ceremonies at both the Camden and Philadelphia plazas. On this date, an estimated 100,000 people were reported to have crossed the world's longest suspension bridge at the time. (Courtesy of the Philadelphia Inquirer.)

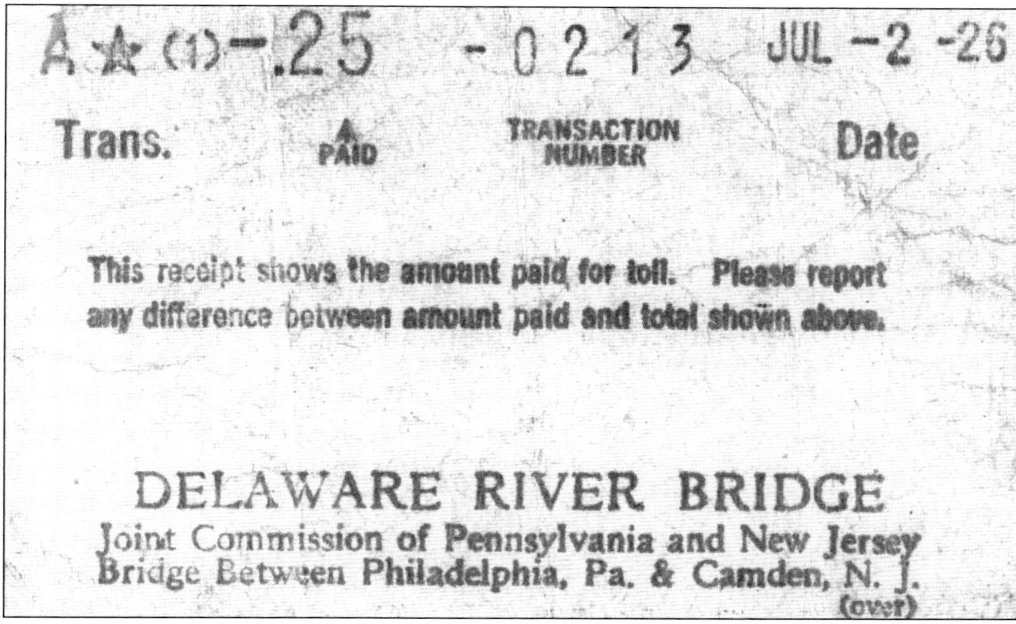
A ★ (1) -.25 -0213 JUL -2 -26

Trans. A PAID TRANSACTION NUMBER Date

This receipt shows the amount paid for toll. Please report any difference between amount paid and total shown above.

DELAWARE RIVER BRIDGE

Joint Commission of Pennsylvania and New Jersey
Bridge Between Philadelphia, Pa. & Camden, N. J.

(over)

At 12:00 a.m. on July 2, 1926, the bridge was opened to vehicular traffic. Some 3,850 vehicles crossed the bridge during the first hour. This image is a receipt from the first day. With a transaction number of 0213, the owner of this receipt crossed the bridge during that first hour. At its opening, the range of bridge tolls included 25¢ for a motor car and 15¢ for a horse and rider, one way.

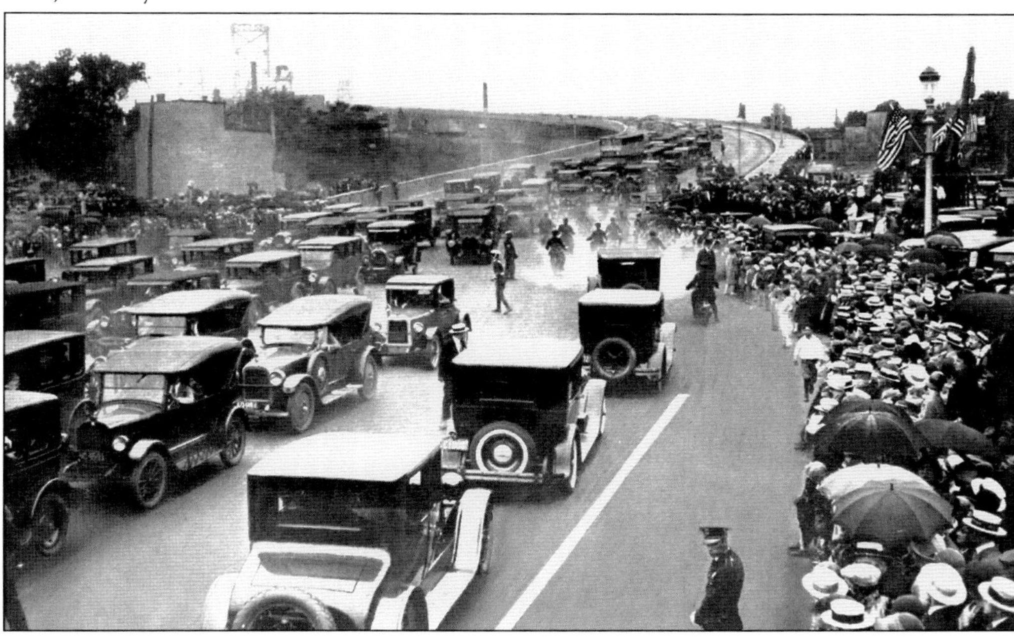

On Monday, July 5, 1926, the bridge was briefly closed to traffic for a dedication ceremony. Pres. Calvin Coolidge was in Philadelphia to attend festivities for the nation's sesquicentennial. To commemorate the bridge's opening, Coolidge planted a tree at the Camden plaza. This image shows the thousands of pedestrians lining the roadway to watch the president's motorcade.

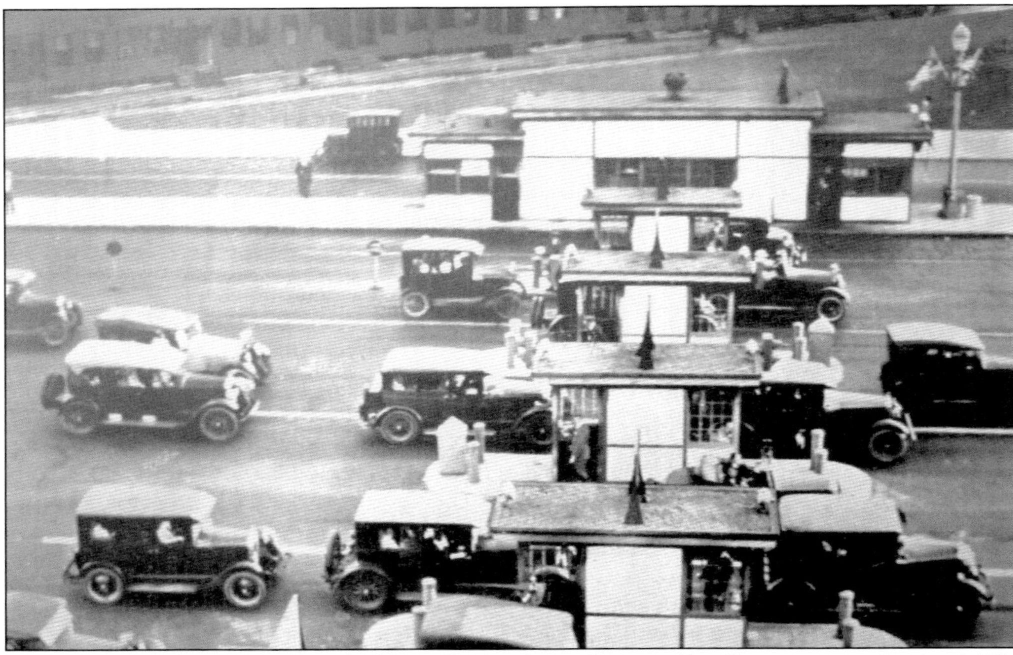

The original toll plaza consisted of six toll islands, with two toll collectors on each island. The center toll lanes were reversible to maximize the number of toll lanes in favor of the direction of heavier traffic. Prior to the bridge's opening, a dispute between the two states regarding the charging of tolls to cross the bridge threatened to halt construction. The dispute was resolved, and construction resumed.

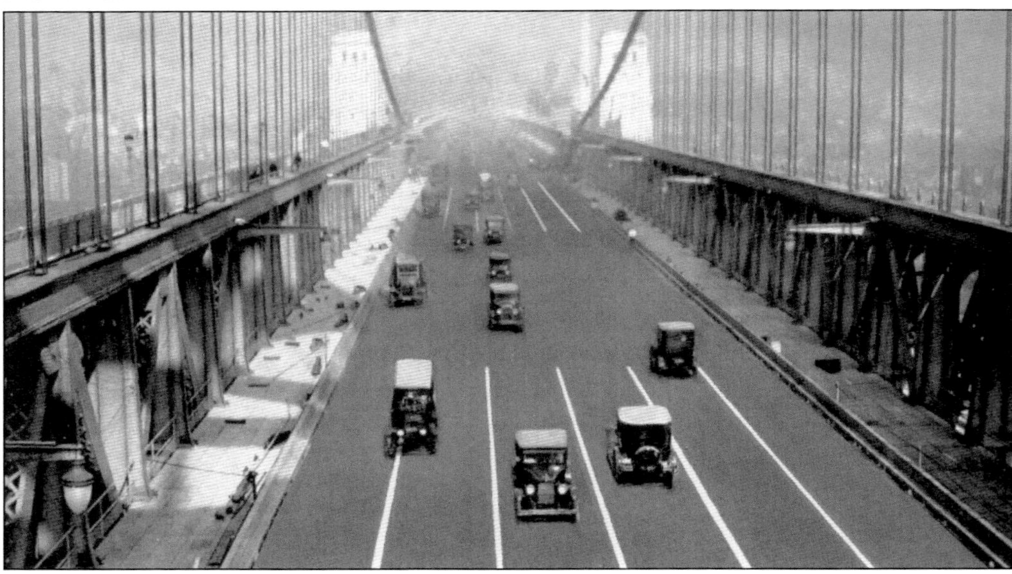

This picture shows bridge conditions during its first year of operation. Although the bridge opened to traffic on July 2, 1926, a few construction projects, such as the main cable wrapping, painting, and electrical and approach work, were not completed until March 1927. In addition to construction wrap-up, maintenance work had begun to address the normal wear and tear on the bridge.

Three

EXPANDING ITS PURPOSE

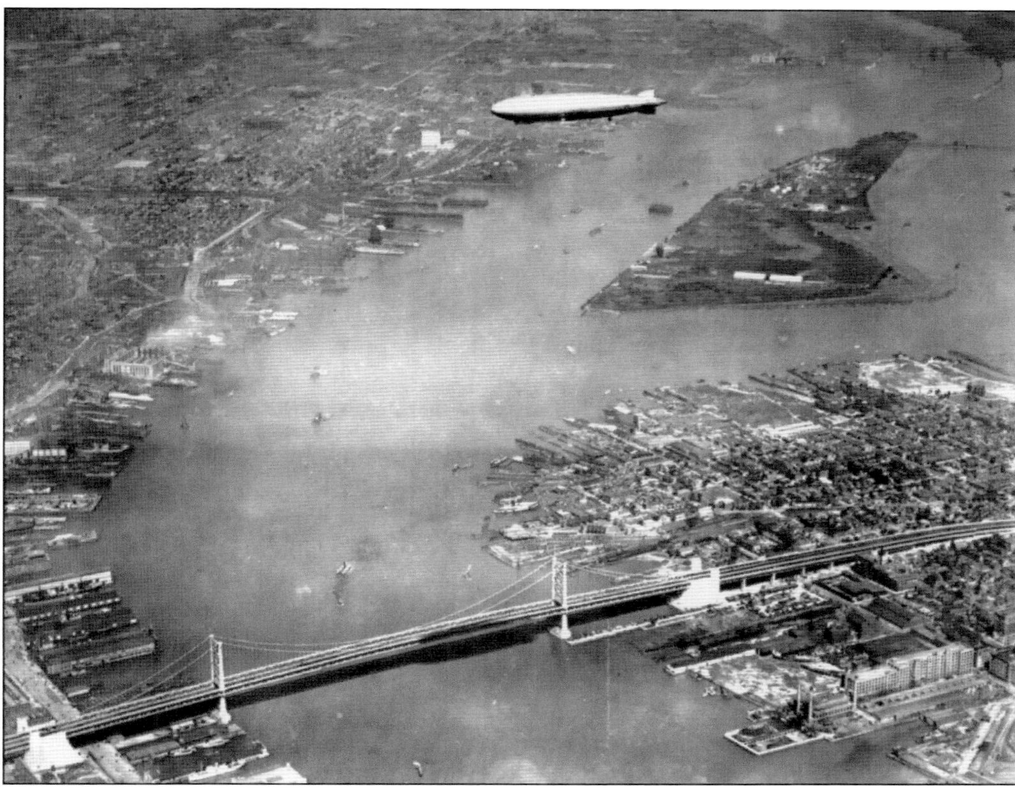

The U.S. Navy dirigible *Los Angeles* flies above the new bridge in 1926, casting its shadow on the main span.

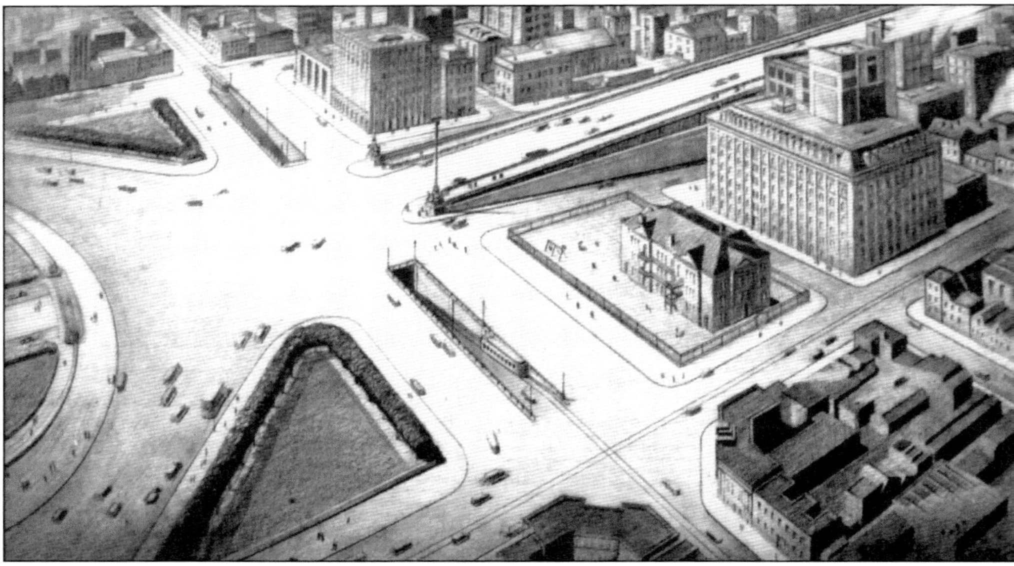

When the bridge opened, traffic on the Philadelphia side had to cross Fifth Street. Provisions were put into place for a traffic light to stop traffic either on the bridge or on Fifth Street. To alleviate traffic back-ups on both the bridge and Fifth Street, a tunnel was constructed to allow vehicles and trolleys to travel between Race and Vine Streets without impacting bridge traffic. This image shows a sketch of the proposed tunnel.

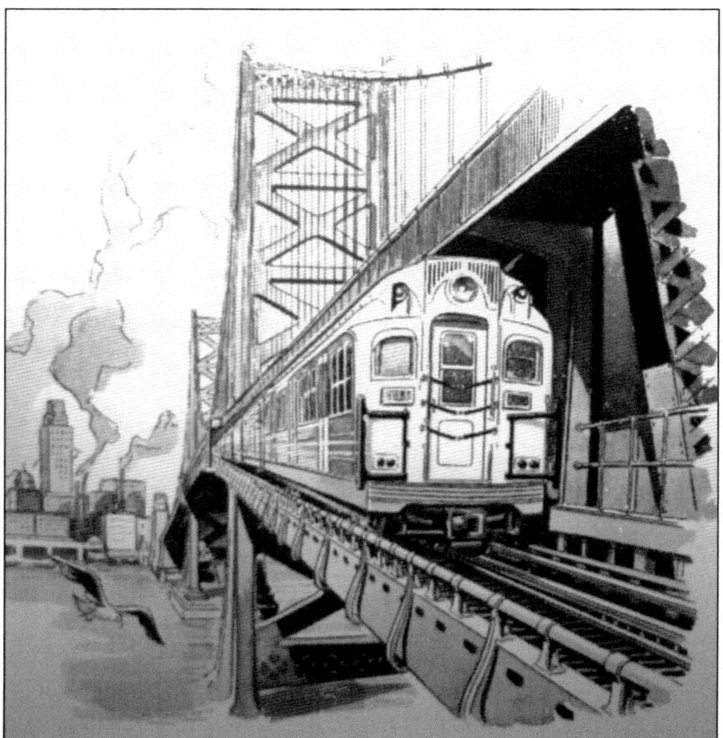

In 1922, the bridge was designed for six lanes of vehicular traffic with four tracks for rapid transit. The Delaware River Joint Commission or DRJC ("Bridge" was dropped from the commission's name in 1931) hoped that the states of New Jersey or Pennsylvania would incorporate these track areas into their respective systems. By 1929, the DRJC began to discuss its options for rapid transit since neither state had stepped forward with a plan.

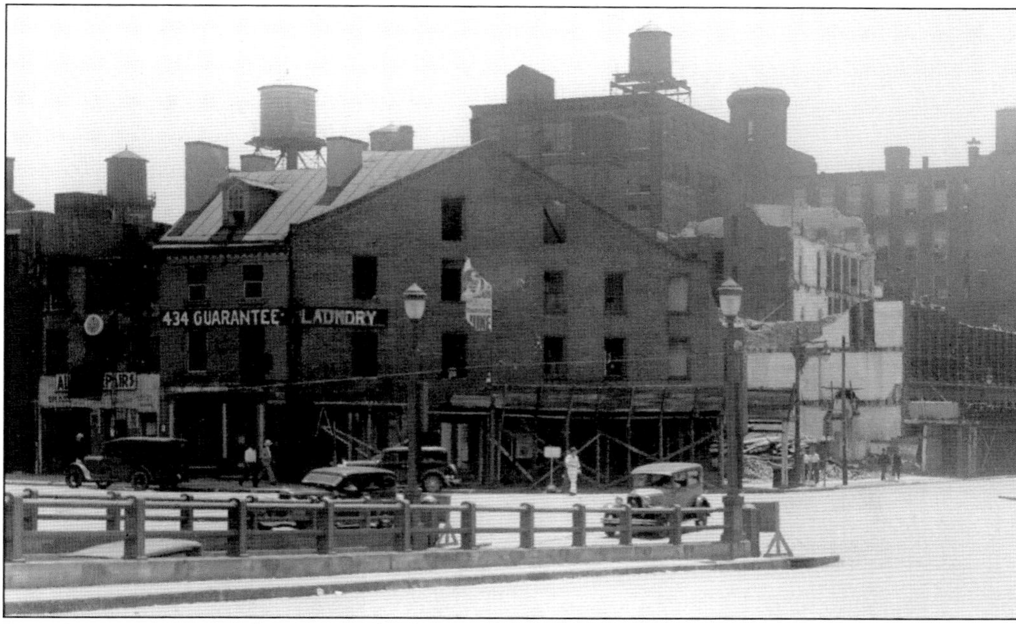

In 1934, construction began on the bridge's rapid-transit high-speed train, the Bridge High Speed Line. The bridge line's alignment necessitated relocation of the Fifth Street tunnel in Philadelphia. The new tunnel would be longer and deeper than the original, constructed in 1930. Here automobiles can be seen traveling north on Fifth Street, entering the south tunnel entrance ramp. In the background, demolition of the buildings on the southeast corner of Fifth and Race Streets has started.

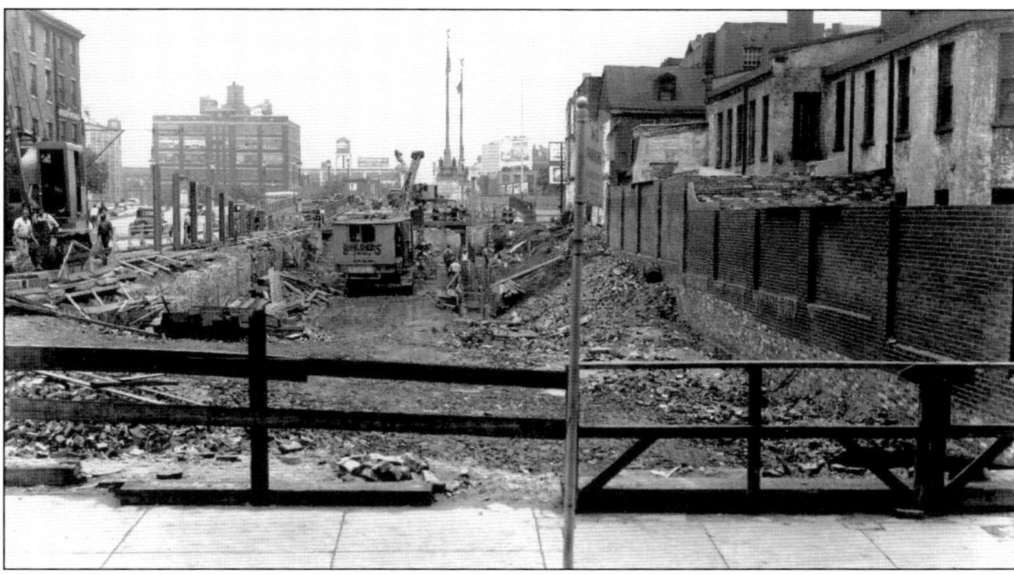

Two months after the prior image, work crews have completed most of the building demolition. Excavation has begun for the new south tunnel entrance, located east of the original tunnel. The buildings on the right are located where the United States Mint is today. In the background, the twin pylons supporting the "Winged Victory" figures are doubling as flagpoles along both sides of the bridge roadway.

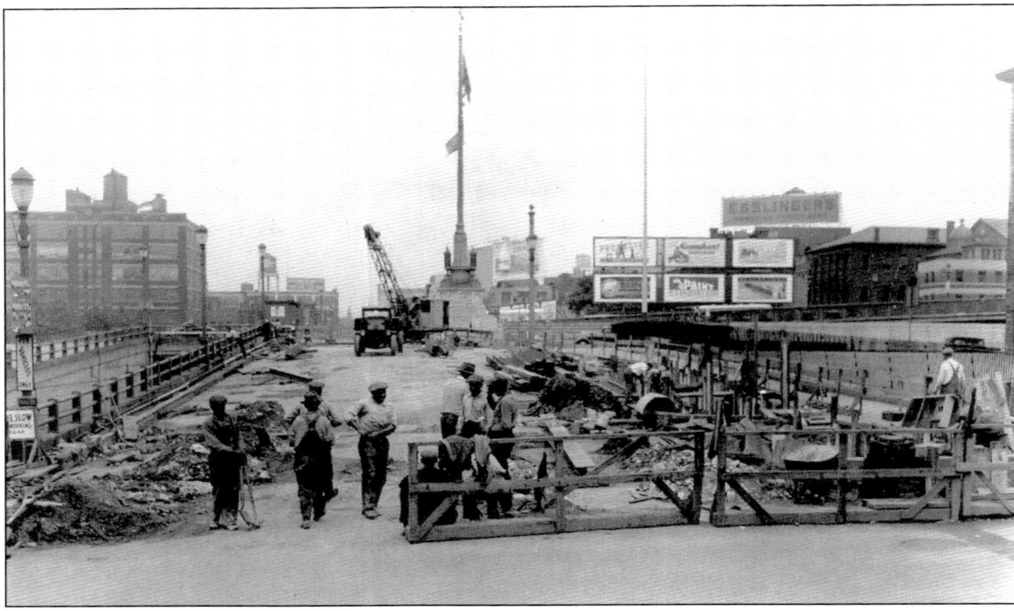

Work crews will begin the excavation process in Philadelphia alongside the existing Fifth Street tunnel (left). The old tunnel remained open during construction. Through this section of Fifth Street, work crews would need to excavate an additional 14 feet below the adjacent tunnel. Once the new tunnel was constructed, a bulkhead was installed at each portal of the old tunnel, and the approach ramps were backfilled.

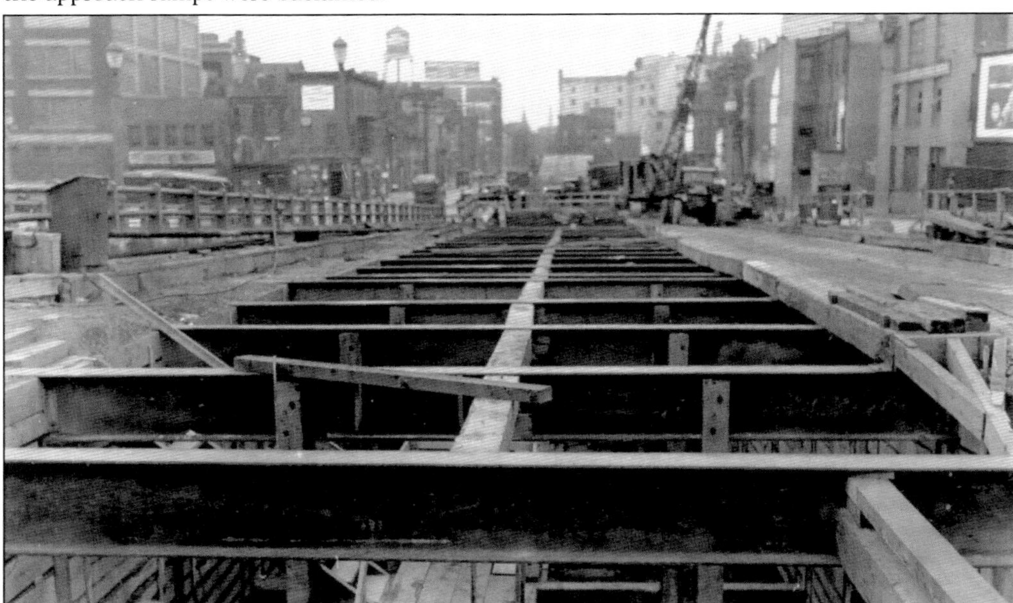

In this view looking north along Fifth Street at the bridge, tunnel excavation is nearing completion. Although the old tunnel (left) ended at Vine Street, the new tunnel continued for another two blocks to Callowhill Street. In addition to serving automobile traffic, trolleys would operate in the new tunnel. At Vine Street, the tunnel would contain a trolley station with a stairway entrance up to street level.

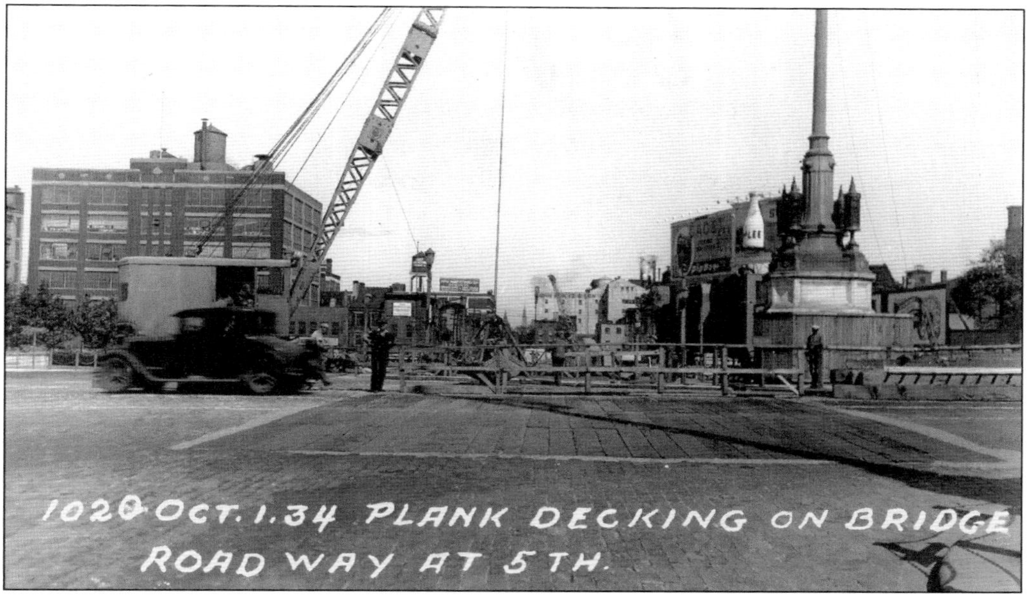

Where the tunnel crosses the end of the bridge, timber planking is used to cover the excavation site. The tunnel planking is in place to minimize disruption to bridge traffic during construction. Supports were installed beneath the temporary surface to bear the weight of traffic crossing the area of excavation.

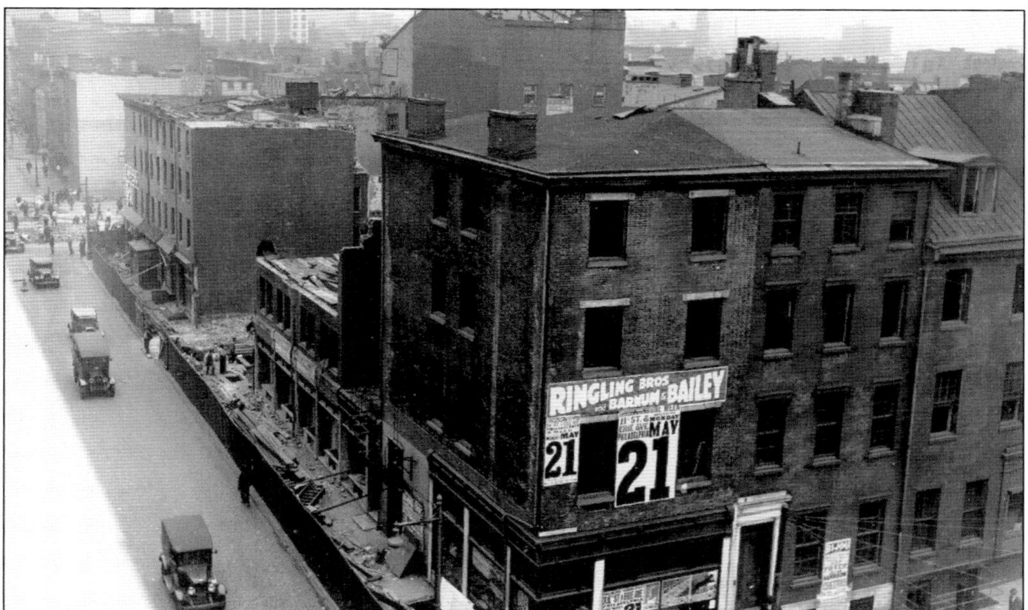

In 1922, the original bridge design plans called for a high-speed tunnel alignment to curve around the bridge's Philadelphia plaza and turn south onto Seventh Street. This alignment was shifted one block to the west, where it would connect with the Philadelphia Rapid Transit's subway on Eighth Street, now called the Ridge Avenue Spur of the Broad Street Subway. As a result of the new alignment, several structures were demolished on Race Street between Franklin and Eighth Streets to make way for the new subway.

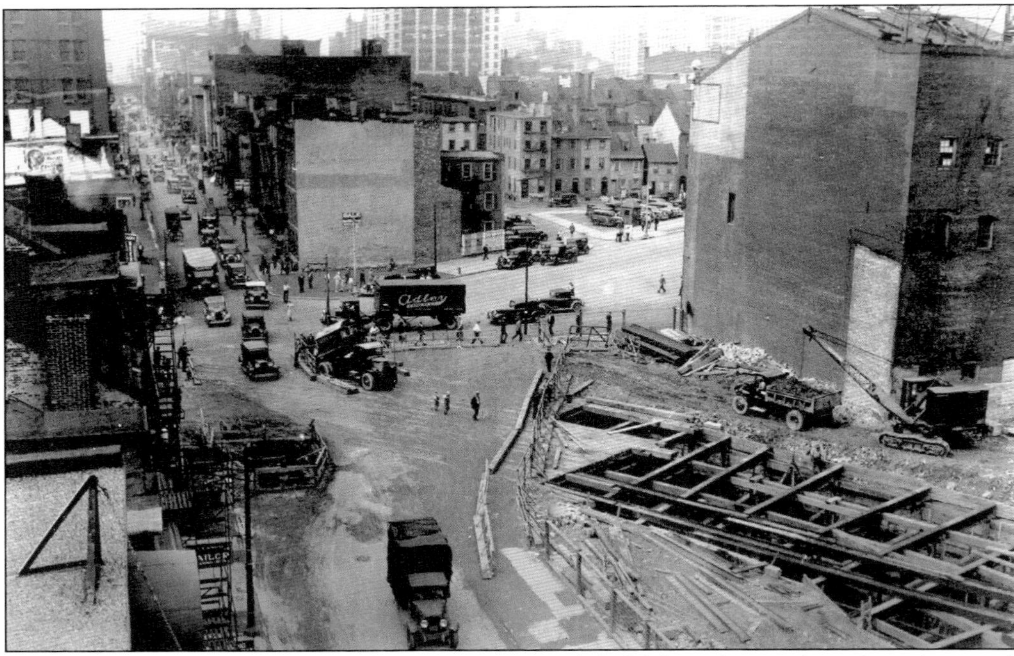

At the intersection of Eighth and Race Streets in Philadelphia, the north and south tunnels turn south to connect with the Eighth Street subway. During subway construction, traffic had to navigate through several detours around the work areas. At this location, the south tunnel will connect with the Eighth Street subway. The north tunnel passes beneath the existing subway and will connect to the subway between Market and Arch Streets.

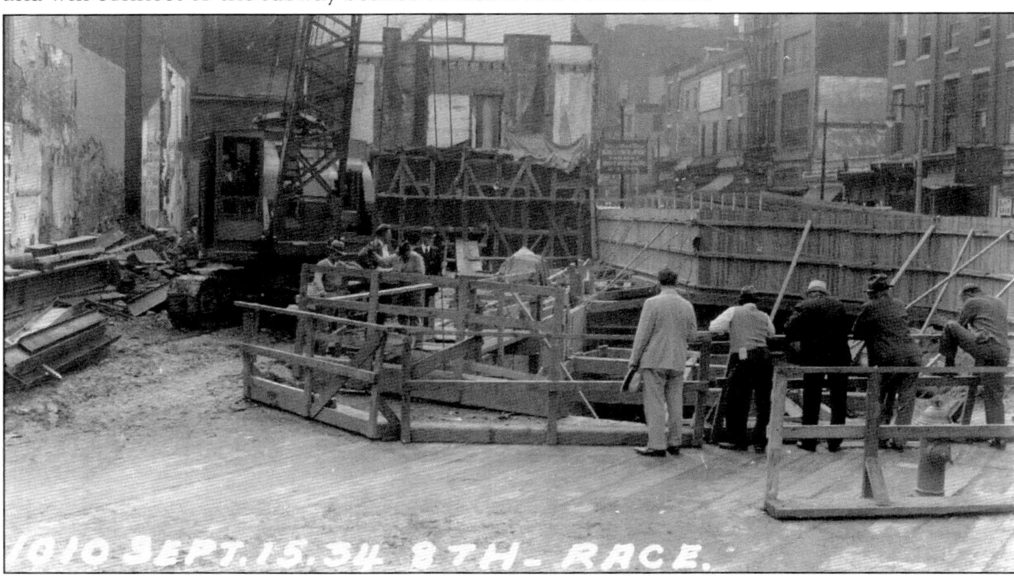

The "sidewalk superintendents" are out in this Saturday scene, overseeing the tunnel excavation on the southeast corner of Eighth and Race Streets. Due to the curvature of the tunnels entering Eighth Street from Franklin Square, the property at this corner had to be demolished. The "superintendents" are standing on the temporary timber roadway that was installed as part of the excavation across the intersection.

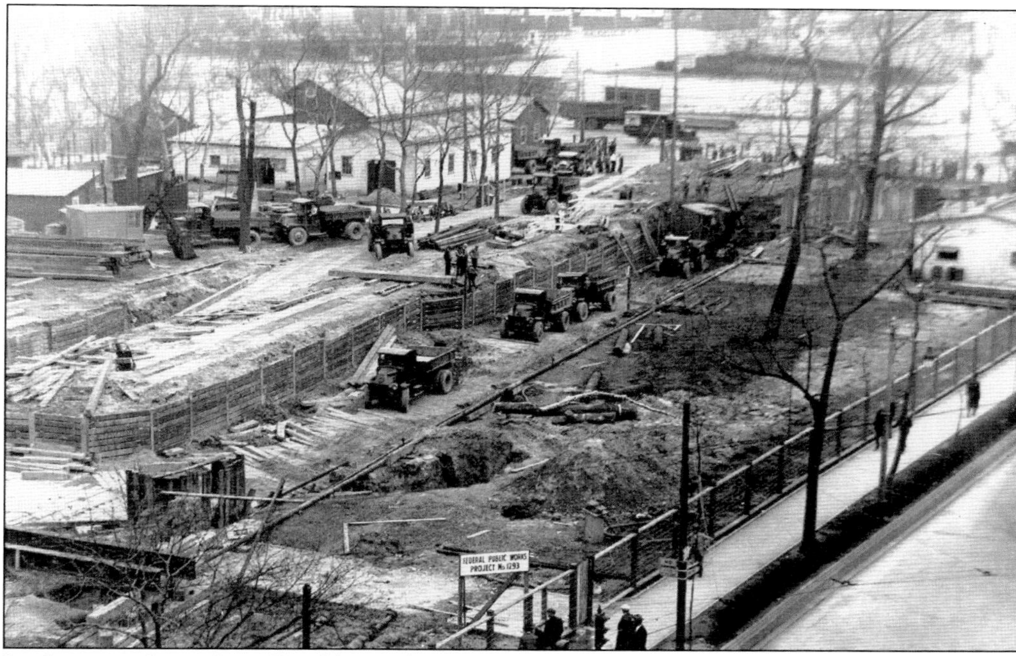

Looking northeast from Franklin and Race Streets in Philadelphia, this view shows the work area within Franklin Square. The alignment of the tunnel through the square would include a station within one of William Penn's original five public squares. The image shows excavation occurring within the south tunnel.

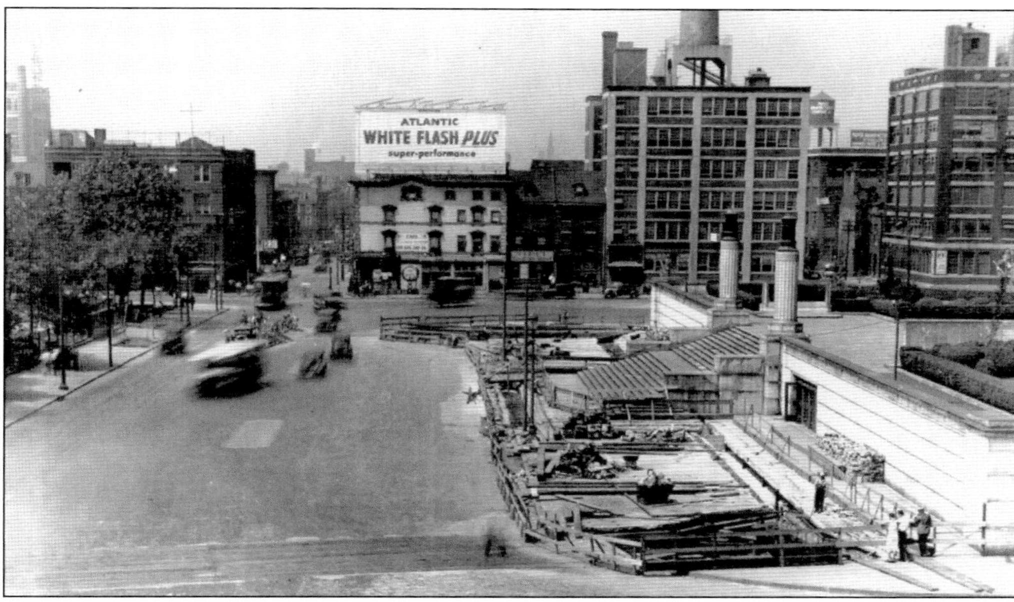

In 1926, construction plans for the Delaware River Bridge contained provisions for the subway, terminating at the Philadelphia plaza (located approximately where the three individuals are standing at lower right). To extend the subway west from the plaza, tunnels for the north and south tracks had to be constructed beneath Sixth Street. Timber planking can be seen on Sixth Street, where excavation is underway for the south tunnel.

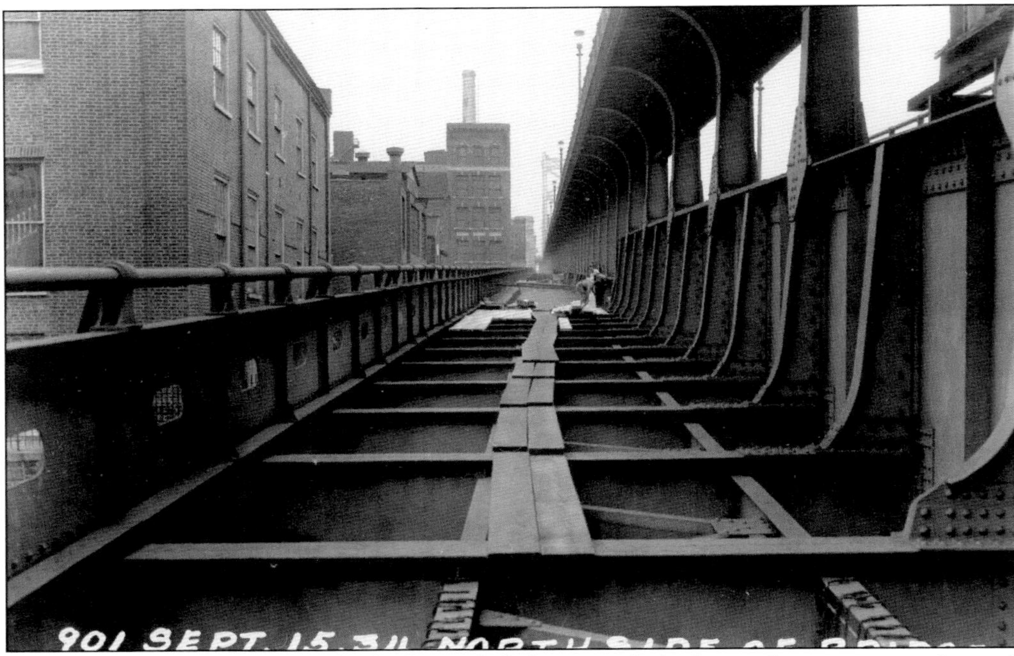

Although the construction of the bridge included an area for high-speed train service, the track areas were not completed when the bridge opened. This view shows the north track where it crosses Fourth Street in Philadelphia. Work crews are preparing the steel for pouring the concrete to create a structure that will support the ballasted track between the anchorage and bridge abutments.

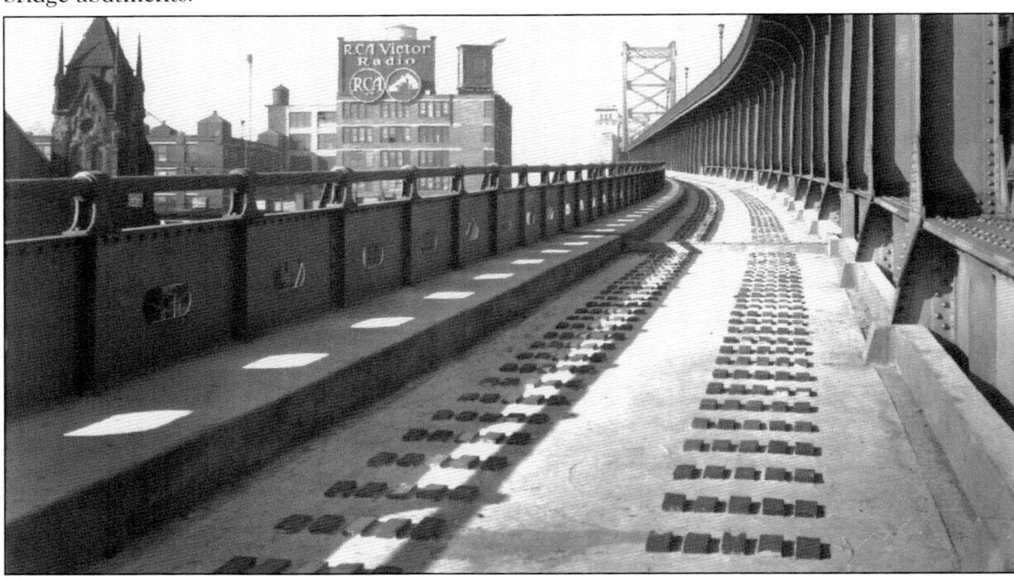

This view of the south track on the Camden approach shows the concrete support slab for the track. Groups of bricks (keystones) set into the concrete every 24 inches assisted with keeping the ballast (stone) and track in position on the steep slopes. The concrete for these approaches remained in service for almost 50 years, until it was removed in the mid-1980s due to deterioration. (Courtesy of Modjeski and Masters, Inc., Mechanicsburg, PA.)

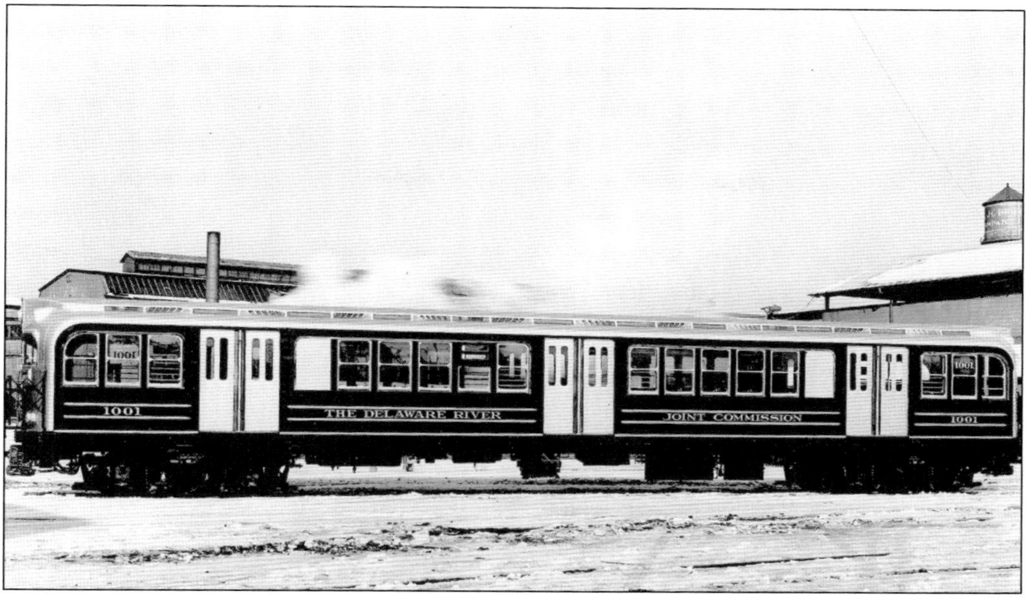

The DRJC purchased 26 transit cars for the Bridge Line operation. The cars were built by the J. G. Brill Company in Philadelphia. Since the cars were to operate on both the Bridge Line and Broad Street Subway, each car was constructed to meet the standard gauge and clearance requirements for both systems. On January 21, 1936, official photographs were taken of the first transit car, No. 1001. A sign in the window on the right side of the bottom picture warns that there is a $25 reward for the arrest and conviction of individuals found tampering with equipment, destroying property, or interfering with the operation of subway elevated trains.

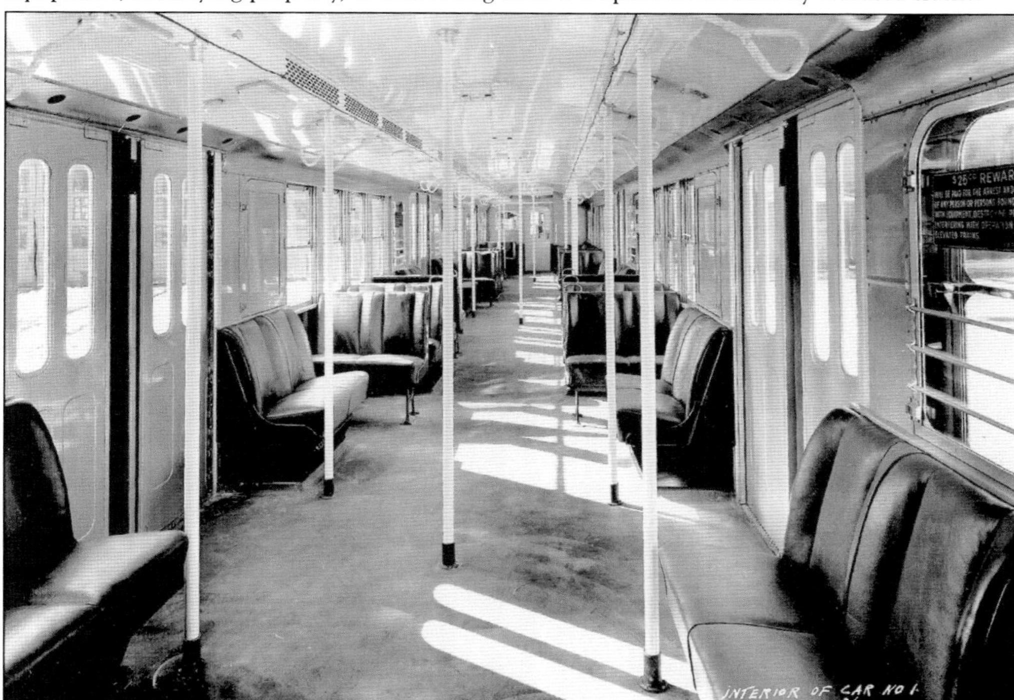

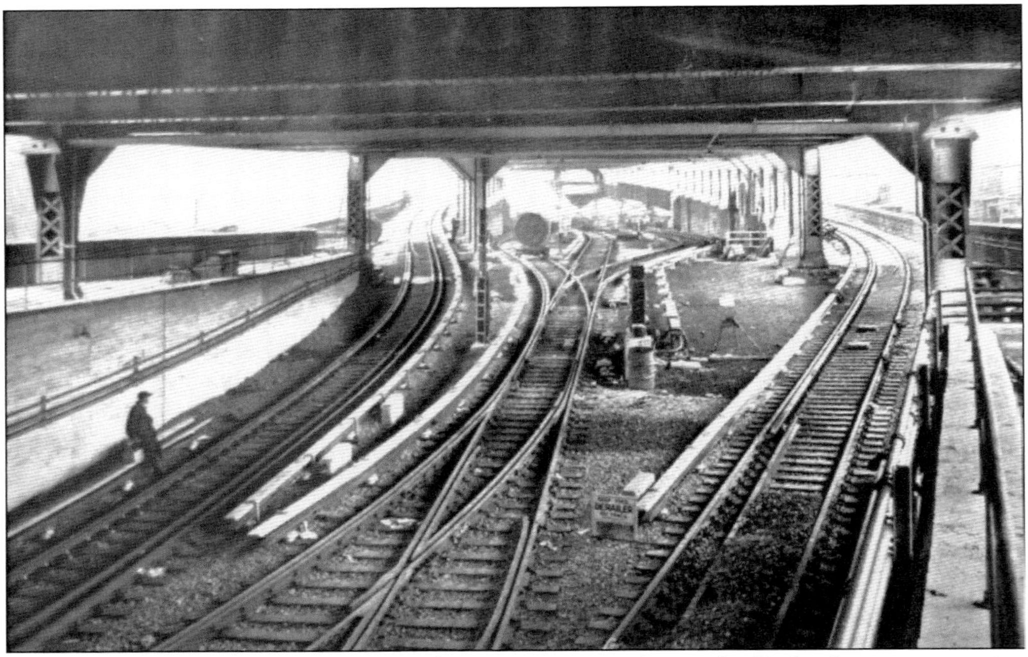

Provisions for a three-track storage yard were incorporated into the Bridge Line construction in Camden. The image above shows the configuration of the tracks leading from the main line tracks to the storage yard. Trains operating on either main line track could enter the yard. Taken two days before the start of operation, the photograph below shows two of the tracks being used to store a few of the 26 transit cars. The center track is being used for maintenance equipment. Bridge Line operation began on June 7, 1936. Although the storage yard still exists beneath the bridge, it is no longer used for passenger vehicles. Today the yard is used for construction equipment.

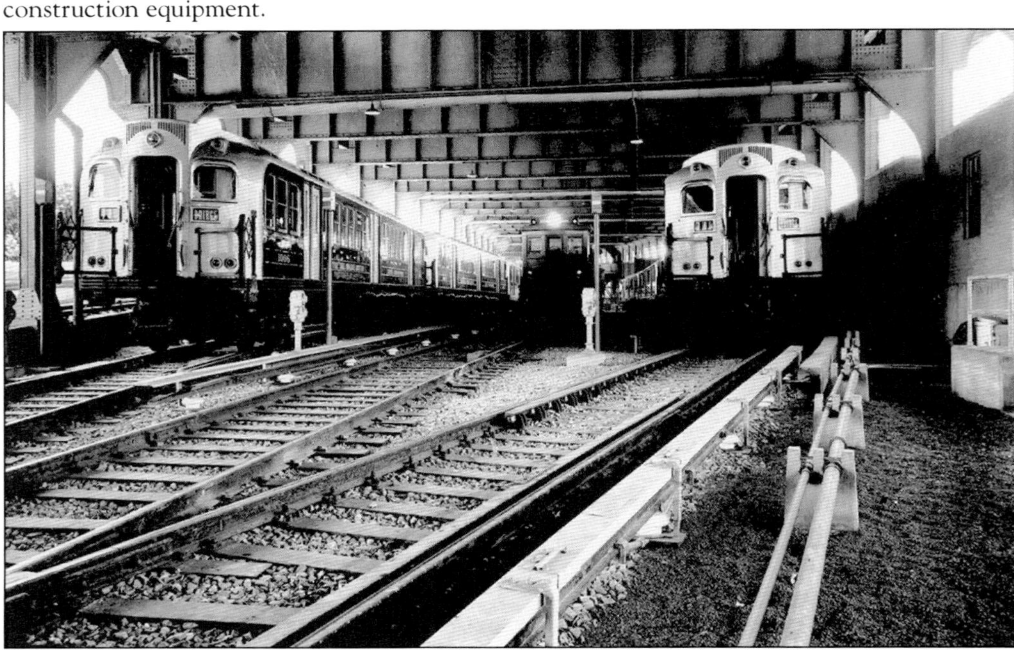

In 1929, the DRJC manufactured 100 directional signs for installation on roadways approaching the bridge. The signs, featuring a stylized bird design, were placed in New Jersey and Pennsylvania to direct motorists toward the bridge. The sign seen here was located at the intersection of Fifth and Chestnut Streets in Philadelphia. In this same year, the bridge lost its designation as the world's longest suspension bridge; Detroit's Ambassador Bridge bested it by 100 feet.

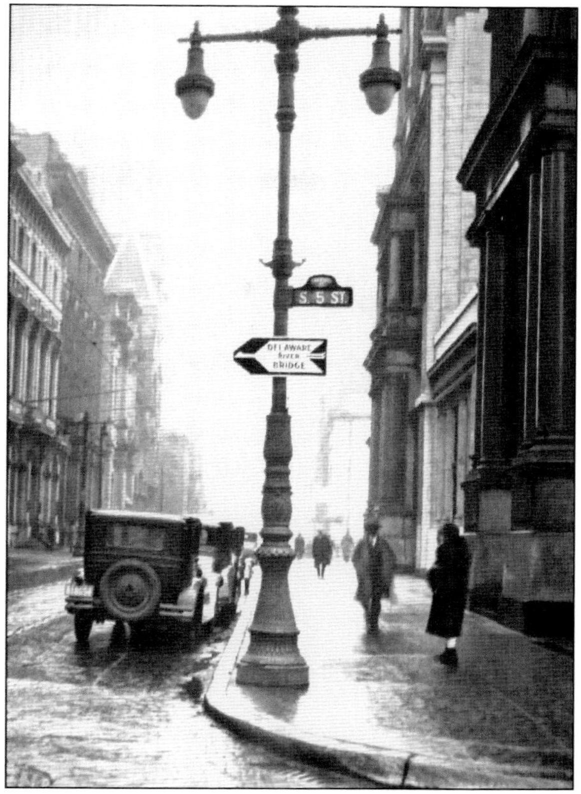

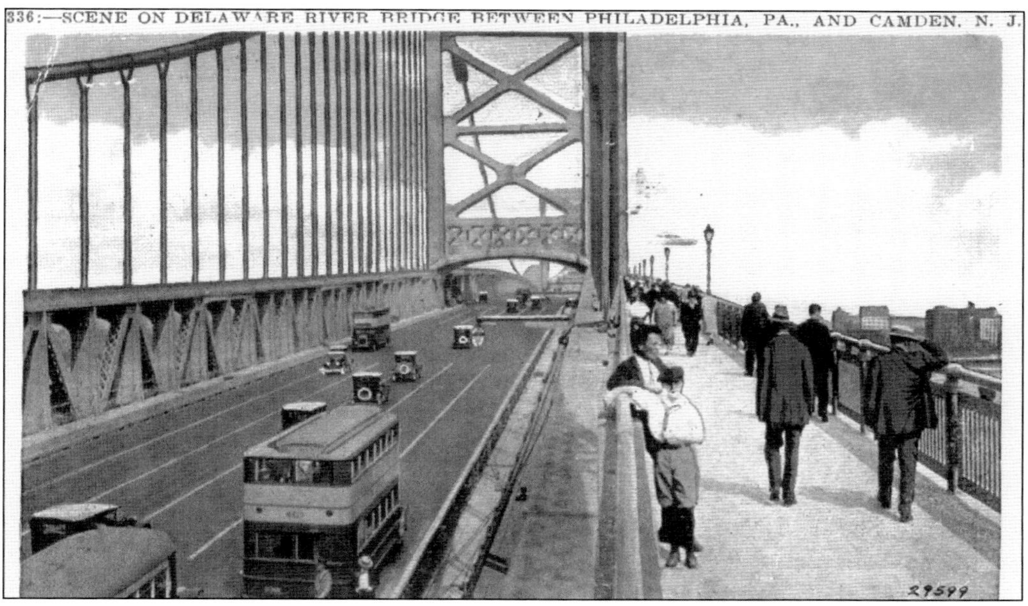

After the bridge was opened to vehicles, pedestrians crossed the bridge via the north and south walkways. In this postcard view looking toward Camden, the south walkway is filled with people strolling across the bridge. The sender of the postcard described the bridge as "High, wide, and handsome—currently enjoying every minute." (Authors' collection.)

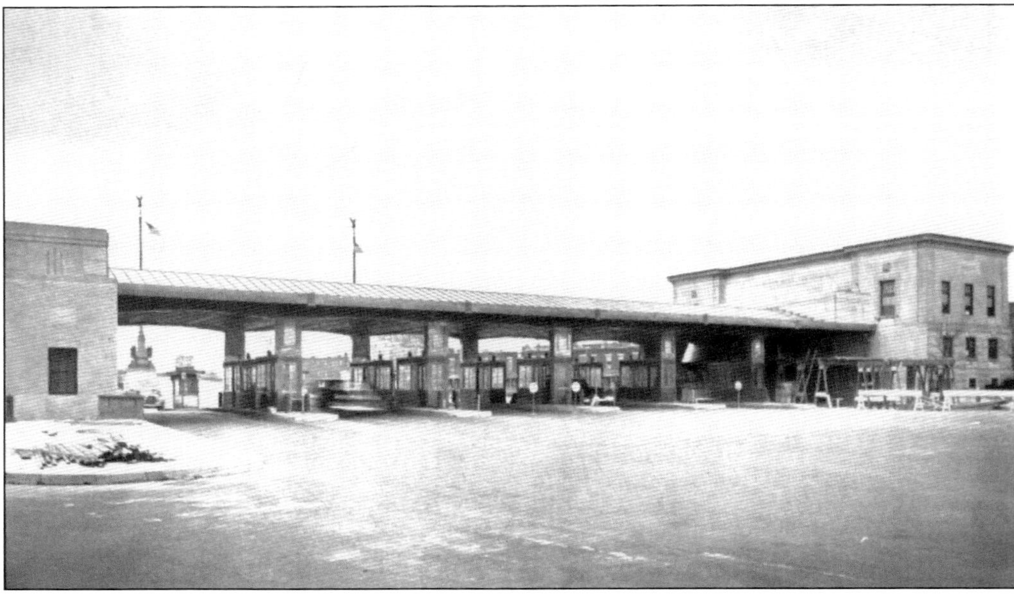

One of the first improvements made to the bridge after its opening was the construction of a new toll plaza east of the existing plaza. The new plaza included a canopy that provided cover to those paying and collecting tolls from each vehicle. The original toll plaza, demolished later, can be seen behind the new plaza. To the right of the new plaza is the bridge's administration building, with offices for those overseeing bridge operations.

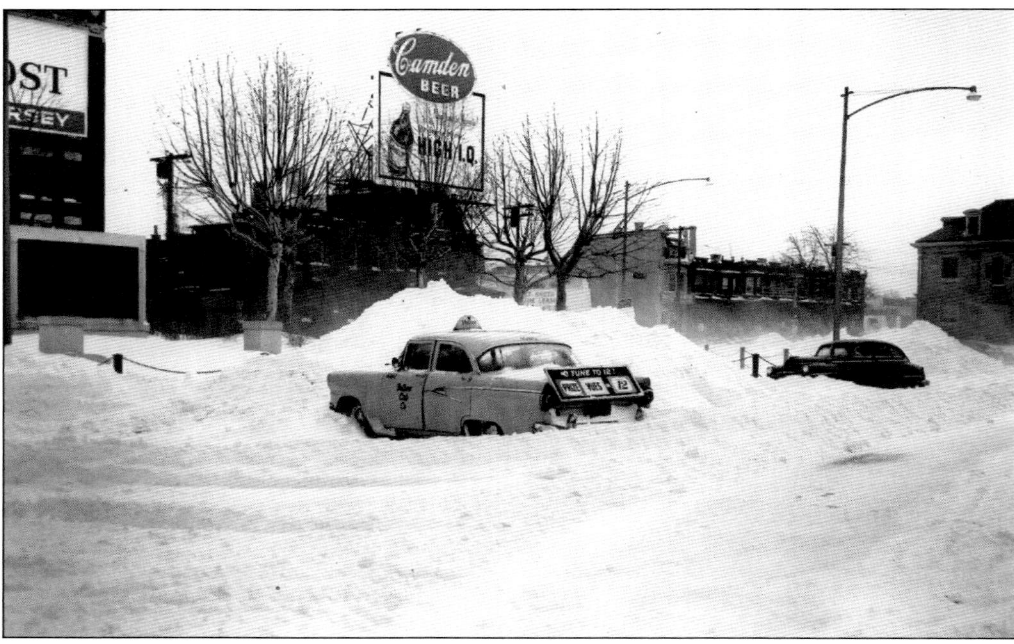

This photograph from February 1958 shows the aftermath of a snowstorm that struck the region. Two vehicles are stuck in snow banks on the east end of the Camden plaza. The taxicab in the foreground features advertisements for a local television station; in the background, there is a billboard for Camden Beer.

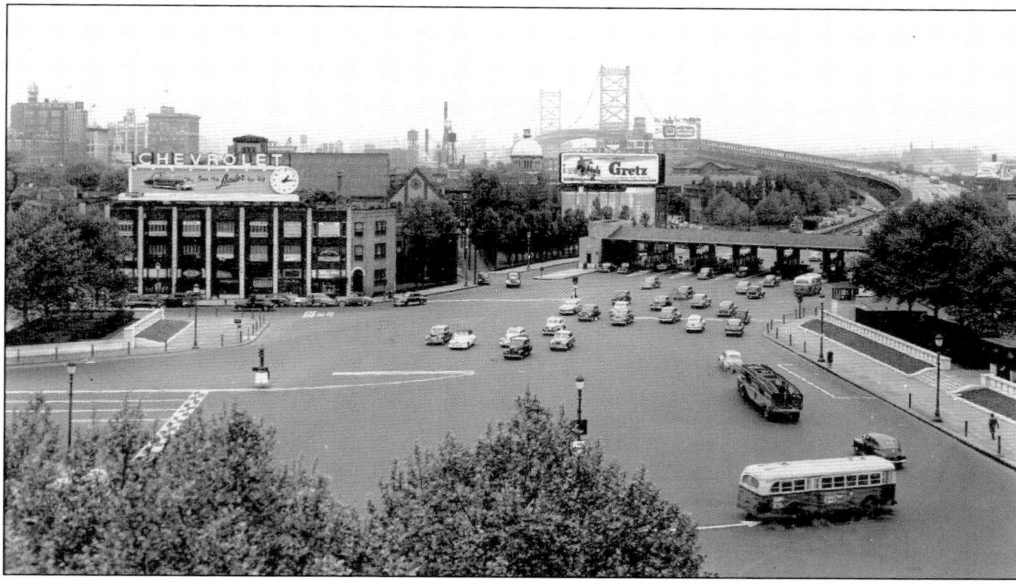

These two views of the Camden plaza were photographed in May 1949 (above) and April 1950 (below). These images show the conditions faced by motorists who were approaching and leaving the bridge. Traffic lights in the plaza center controlled movements throughout the plaza. A police officer working in the small building to the right of the bus entering the toll plaza (above image) controlled the lights to keep traffic moving efficiently. Bus shelters were constructed in the 1930s on both the north (above image, extreme right center) and south (below image, center) sides of the plaza for passengers waiting to board buses heading to, or away from, the bridge.

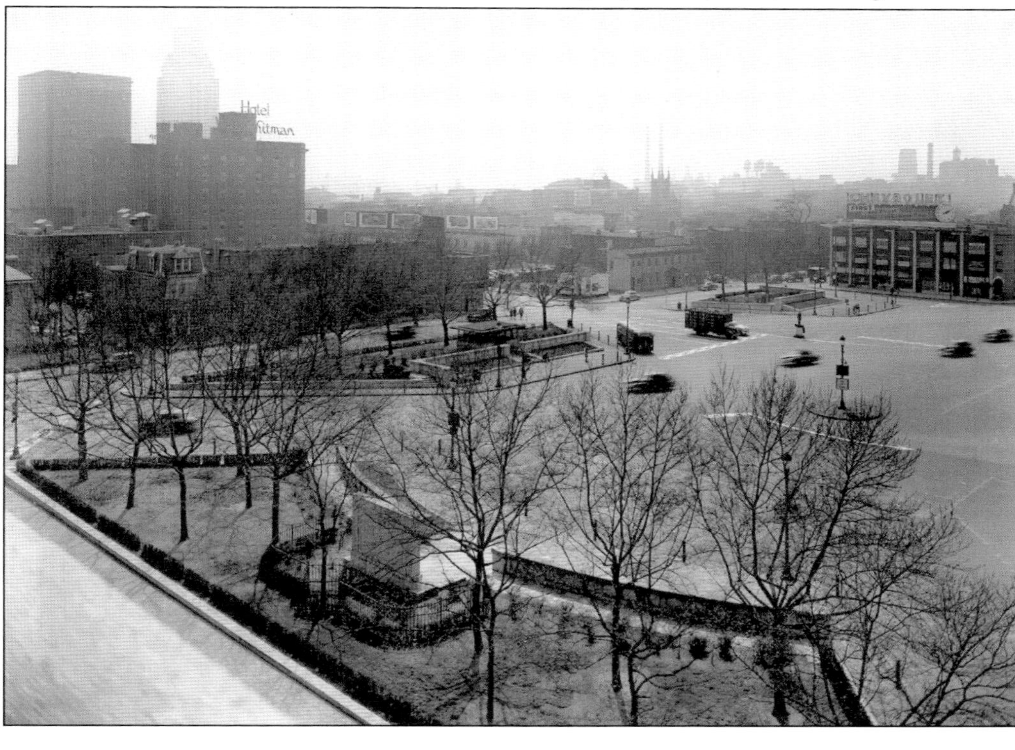

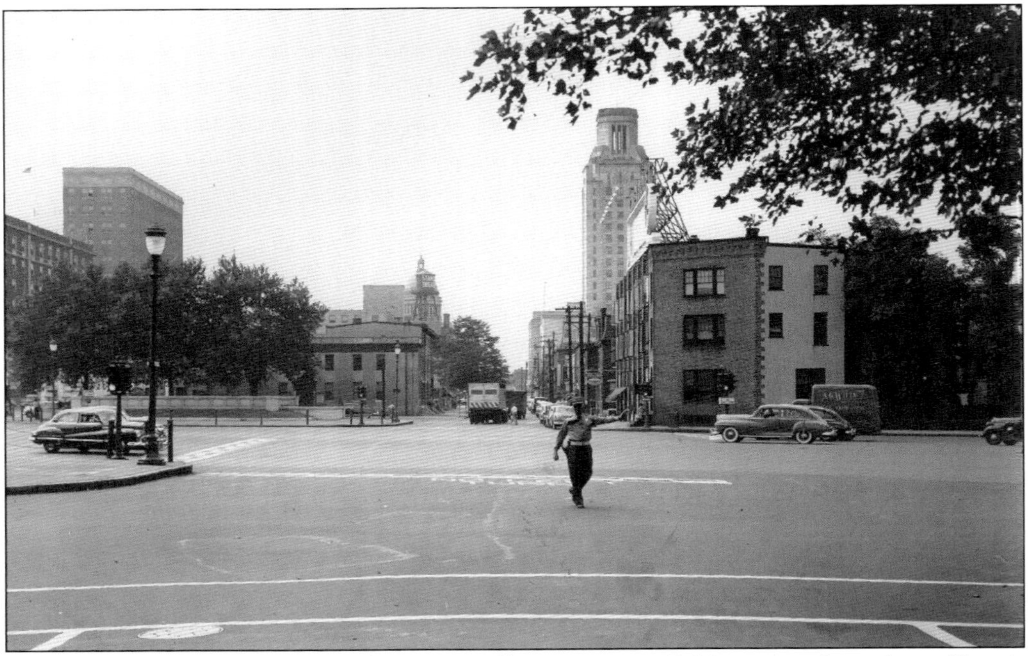

According to the clock on Camden City Hall's tower, it is 12:30 p.m., and a police officer is directing midday traffic in September 1949. This scene, along with the ability to drive south on Sixth Street, was greatly altered when a new, larger toll plaza was constructed three years later.

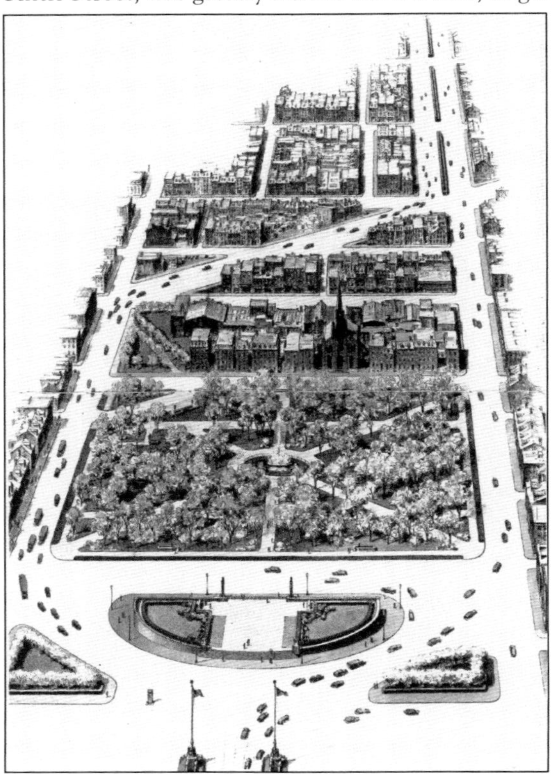

One reason the bridge was constructed with a terminus at Sixth Street between Race and Vine Streets was the capability of Philadelphia's local roads in handling entering and exiting bridge traffic. In 1944, when vehicular volume began to exceed these roads' capacity, city officials decided to widen Vine and Race Streets. This sketch shows the proposed widening and improvements, which included Ridge Avenue. Work on widening the local roads was completed in 1951.

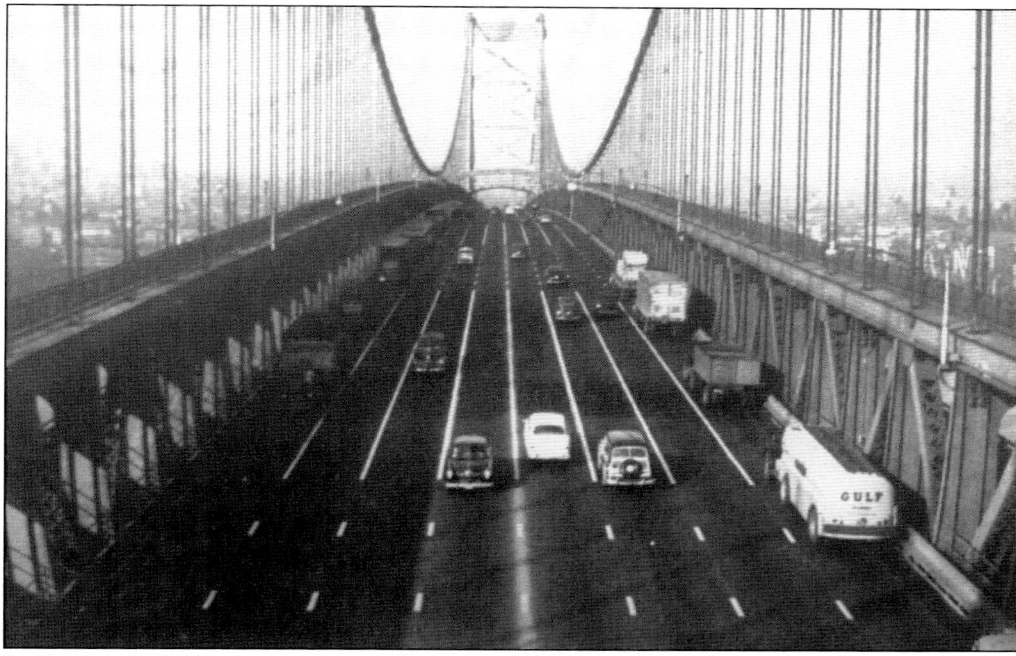

In 1944, the DRJC began to consider elimination of the never-used streetcar track areas to increase roadway capacity. Work on widening the roadway did not start until 1949, when improvements to the Philadelphia street system were undertaken. By 1950, the roadway was widened from six to eight lanes. Four casualties of the widening were the "Winged Victory" angels, which were removed and stored at the Philadelphia plaza for more than 45 years.

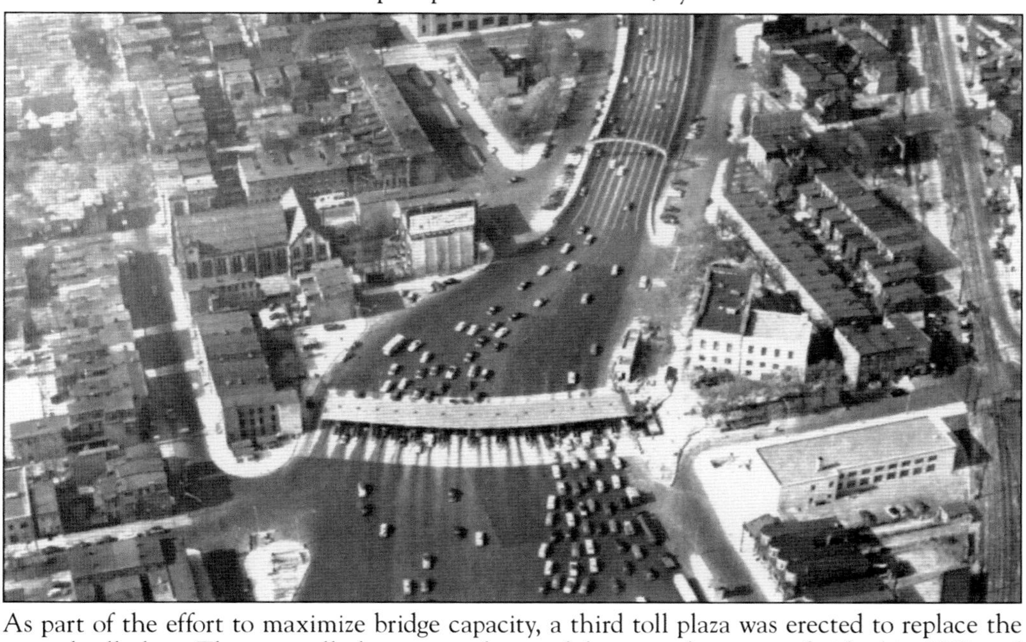

As part of the effort to maximize bridge capacity, a third toll plaza was erected to replace the second toll plaza. The new toll plaza, erected east of the second one, was finished in 1952 and featured two-and-a-half toll lanes for every bridge lane, a ratio thought to be sufficient in meeting the ever-increasing traffic volumes.

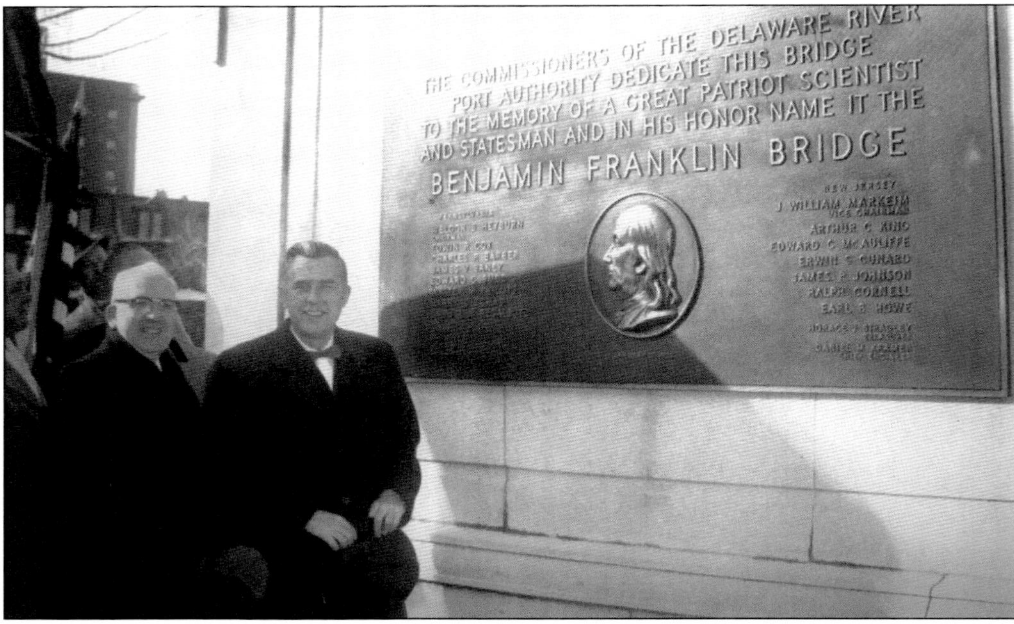

Some 29 years and 20 days after the date of its opening to traffic, the first bridge over the Delaware River connecting Philadelphia and Camden was officially renamed the Benjamin Franklin Bridge. The name selection was based on a recommendation by a Special Committee on Bridge Names. This picture of New Jersey governor Robert Meyner (right) and vice chairman J. William Markheim shows them at a dedication ceremony on January 17, 1956, Benjamin Franklin's 250th birthday anniversary.

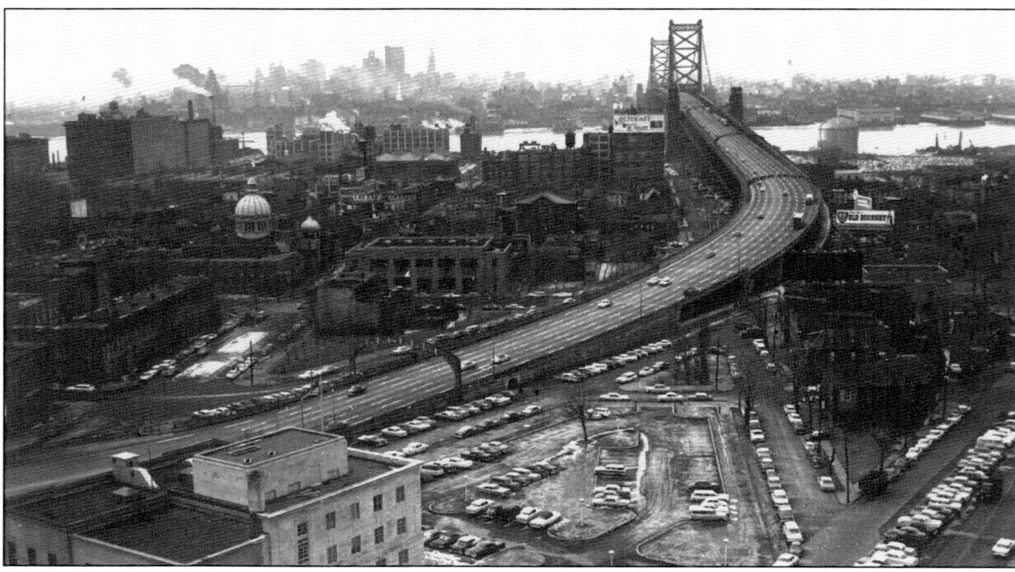

This January 1963 view looking from Camden toward Philadelphia shows the bridge's administration building (lower left corner) and maintenance building (left of middle center). This quiet winter scene reflects a time just prior to the largest expansion phase of the DRPA. In a matter of 10 years, the number of bridge facilities will double, and the Bridge Line will be expanded into a longer, modern, first-of-its-kind transit system.

Four

Additional Crossings

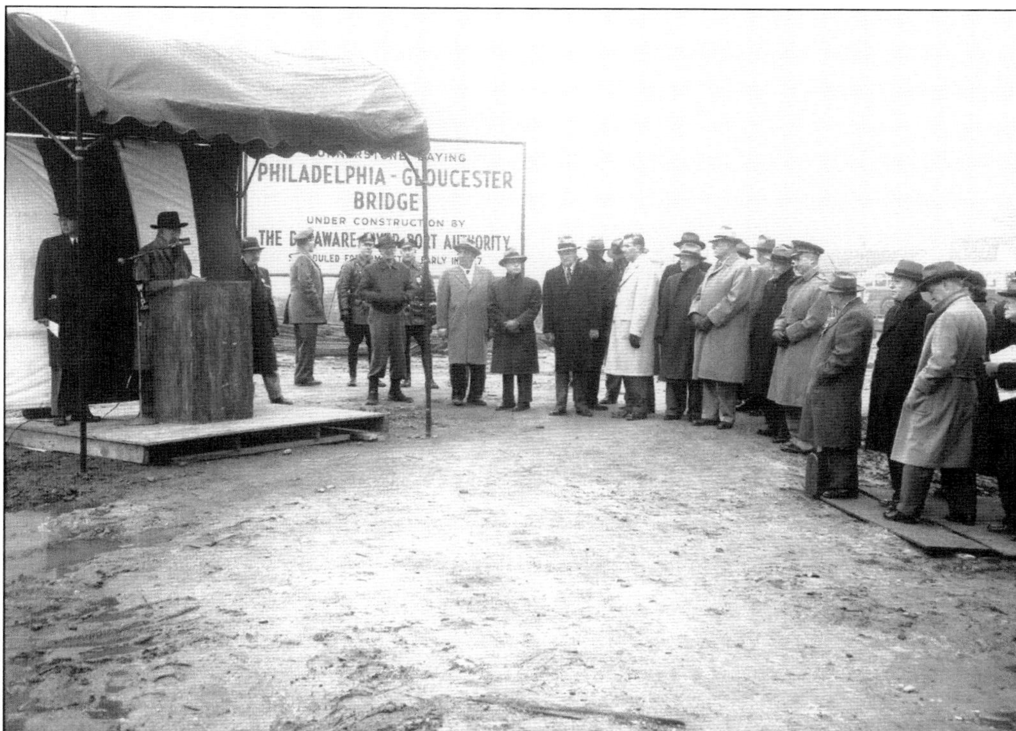

Although attempts were made for 20 years to increase capacity on the Benjamin Franklin Bridge, it was inevitable that a second bridge would be needed. In 1947, the DRJC began to consider a second crossing. The site chosen was between Gloucester City, New Jersey, and Packer Avenue in Philadelphia. The commissioners from the DRJC, renamed the Delaware River Port Authority (DRPA) in 1952, are seen here at the laying of the anchorage cornerstone for the forthcoming Walt Whitman Bridge.

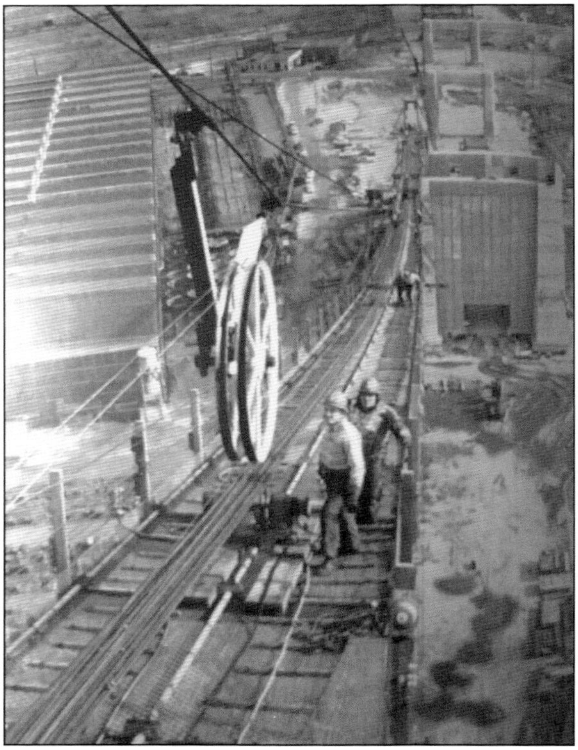

Workers on the Gloucester side span stand aside as the cable wheel passes on its way to the Gloucester anchorage. Although the construction methods were the same as those used in the 1920s, the use of two cable wheels allowed two pairs of wires to be pulled with each pass instead of the single wire pulled on the Benjamin Franklin Bridge. The footbridge on which the workers are standing is made of a chain-link fabric rather than the wooden platforms used in 1926.

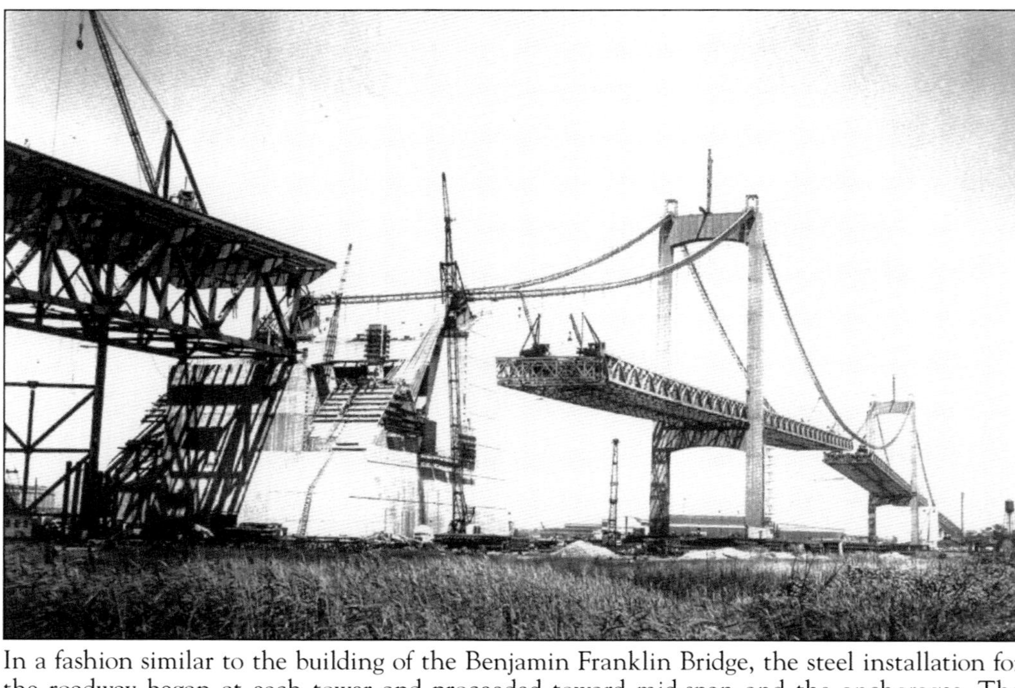

In a fashion similar to the building of the Benjamin Franklin Bridge, the steel installation for the roadway began at each tower and proceeded toward mid-span and the anchorages. The Walt Whitman Bridge has a longer main span and higher clearance above the river than its upstream companion.

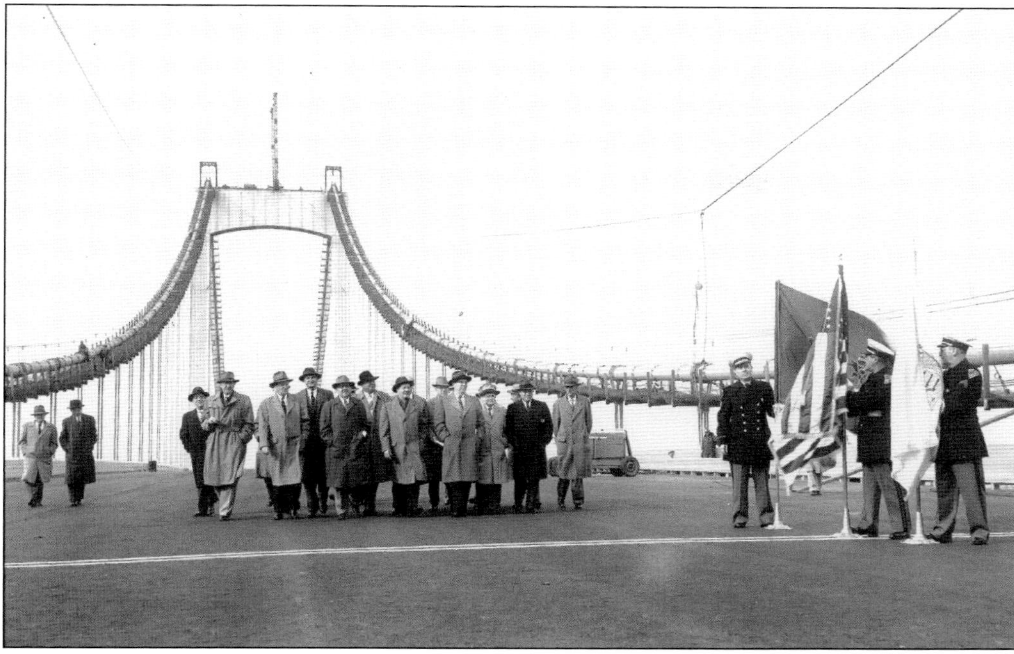

On November 21, 1956, the DRPA commissioners made the first official crossing of the new bridge. As the commissioners approach mid-span, a double white line can be seen painted across the roadway to denote the Pennsylvania (far side) and New Jersey (near side) state boundary.

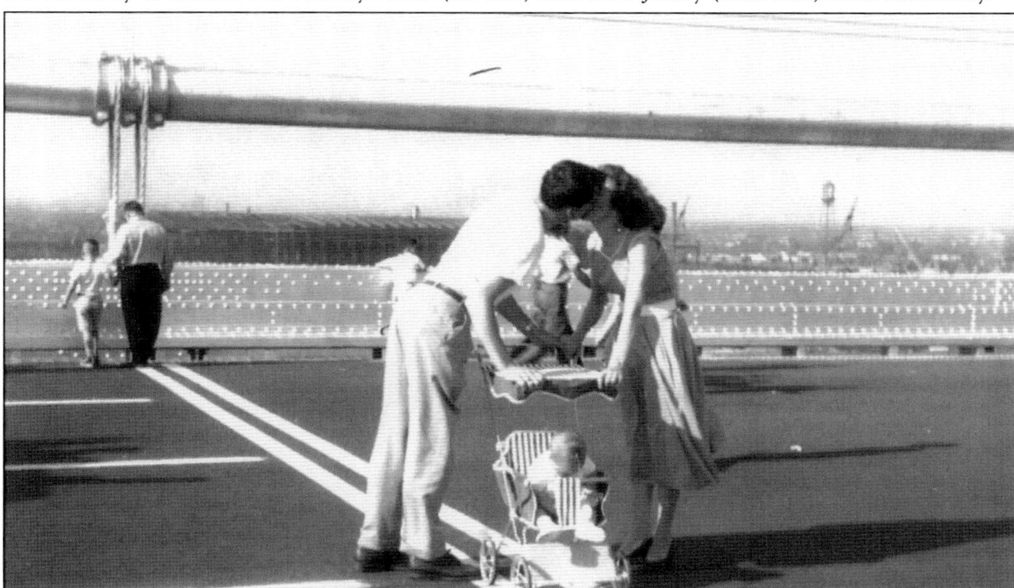

On May 15, 1957, the Walt Whitman Bridge was formally opened with a ceremony and speeches by representatives from Pennsylvania, New Jersey, the DRPA, and engineering firms involved with bridge construction. Between noon and 6:00 p.m., the bridge was open to pedestrians. In this image, Gloucester City resident Dorothy Stanaitis and her husband, Raymond, share a moment where their two favorite states meet. Their son Jim sits in the stroller. (Courtesy of Dorothy Stanaitis.)

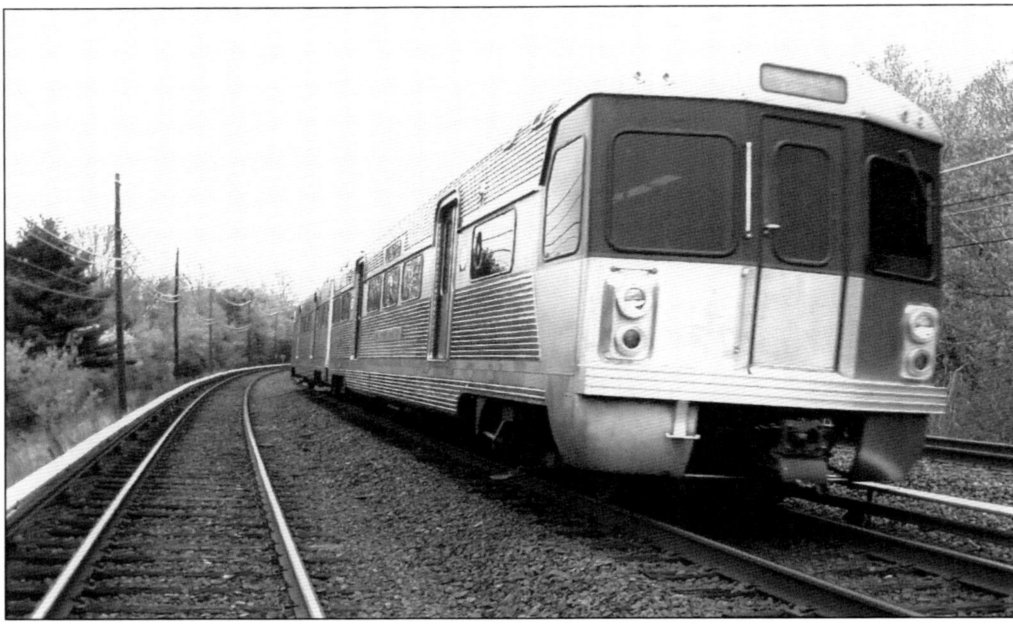

In 1956 and again in 1961, studies were performed to evaluate expanding the Bridge Line operation. Both of the studies examined extending train service further into New Jersey. The 1961 study recommended using the Bridge Line's infrastructure and extending service to Lindenwold, New Jersey. The Port Authority Transit Corporation (PATCO) would operate the "new" high-speed line and feature the stainless steel cars seen in this image. (Authors' collection.)

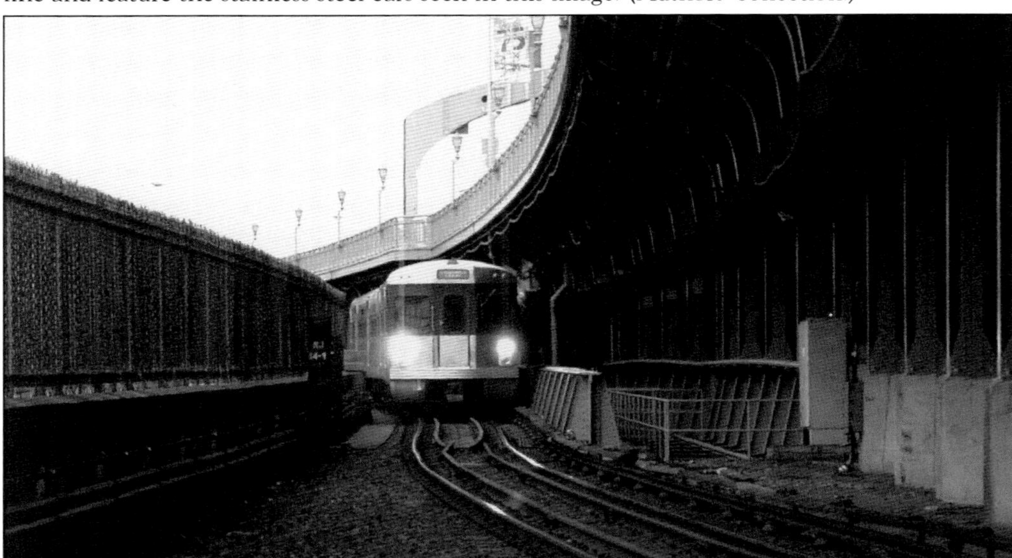

In 2001, an eastbound train descends from the bridge to enter the Camden subway. This image, west of the first image on page 78, shows the alignment of both rapid-transit track areas as originally constructed in 1926 and the modifications performed in 1936. Although the high-speed transit service alignment was originally planned to divert off the bridge at this location, the 1936 construction moved the track alignment beneath the bridge, blocking potential streetcar access. (Authors' collection.)

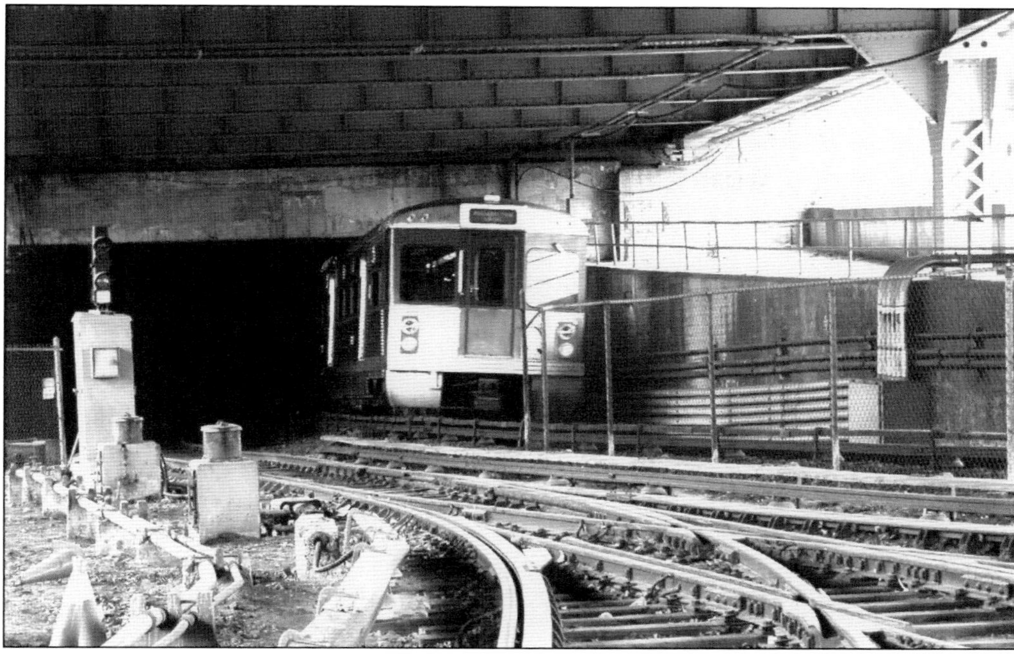

As an eastbound train enters the Camden tunnel, the tracks in the foreground lead to the three-track storage yard seen in the first image on page 106. Although the transit equipment has changed, many of the track components seen in those images were still in place when this picture was taken 65 years later. (Authors' collection.)

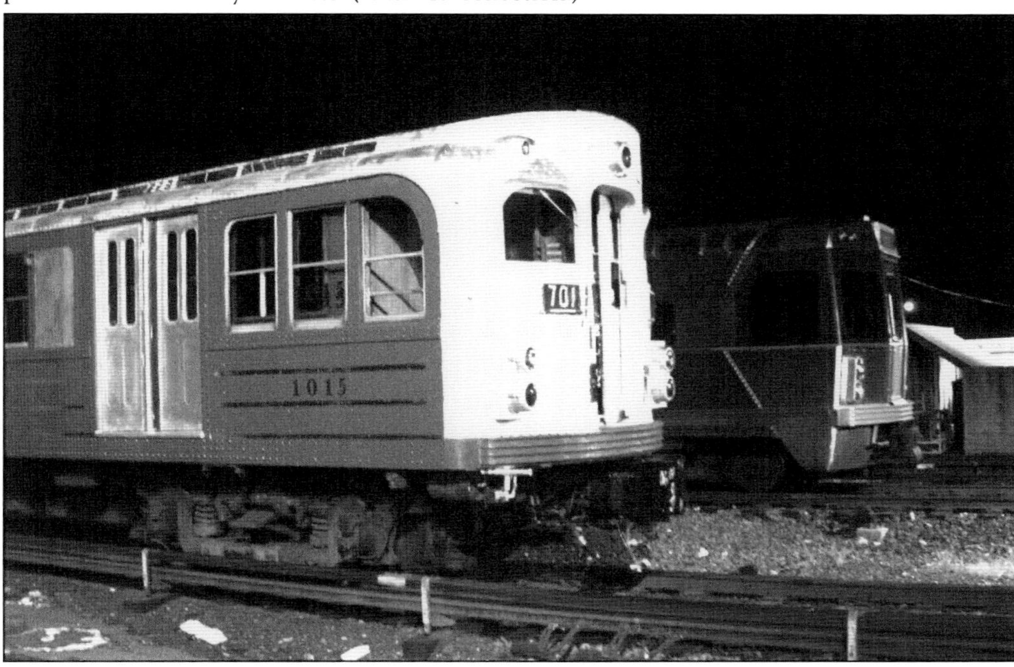

In 2001, two original Bridge Line transit cars were repainted to commemorate the 75th anniversary of the Benjamin Franklin Bridge. Seen here at the Lindenwold shop, Bridge Line car (original No. 1015) sits next to a modern PATCO car. (Authors' collection.)

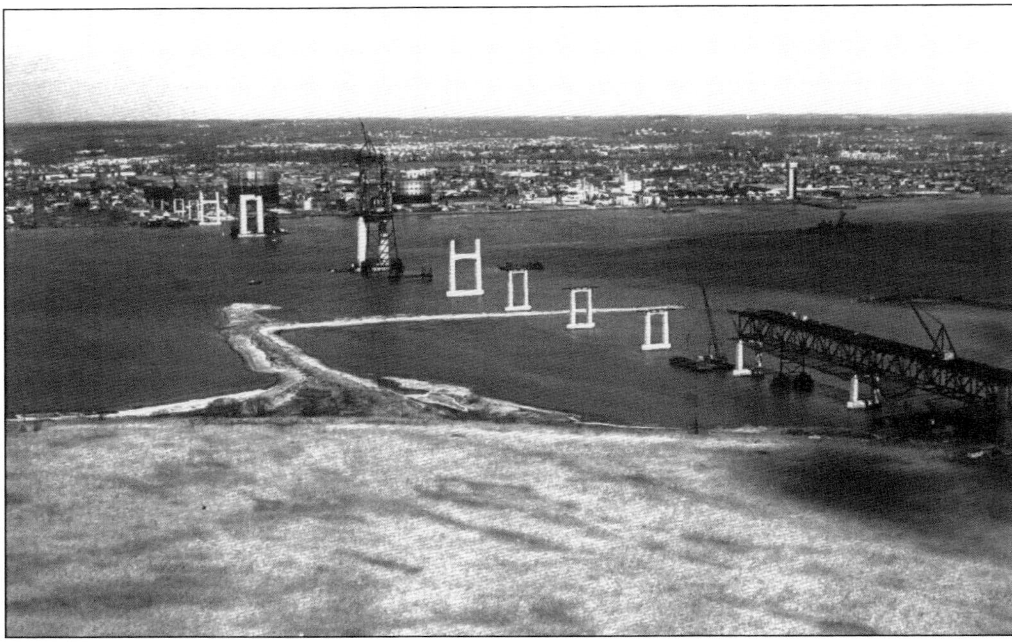

In 1969, the DRPA began construction on the Commodore Barry Bridge between Bridgeport, New Jersey and Chester, Pennsylvania. This cantilever-truss-style bridge was the first non-suspension bridge for the DRPA. Due to its location downstream of the Philadelphia Navy Base, this bridge features the highest clearance above the river for any of the four DRPA bridges.

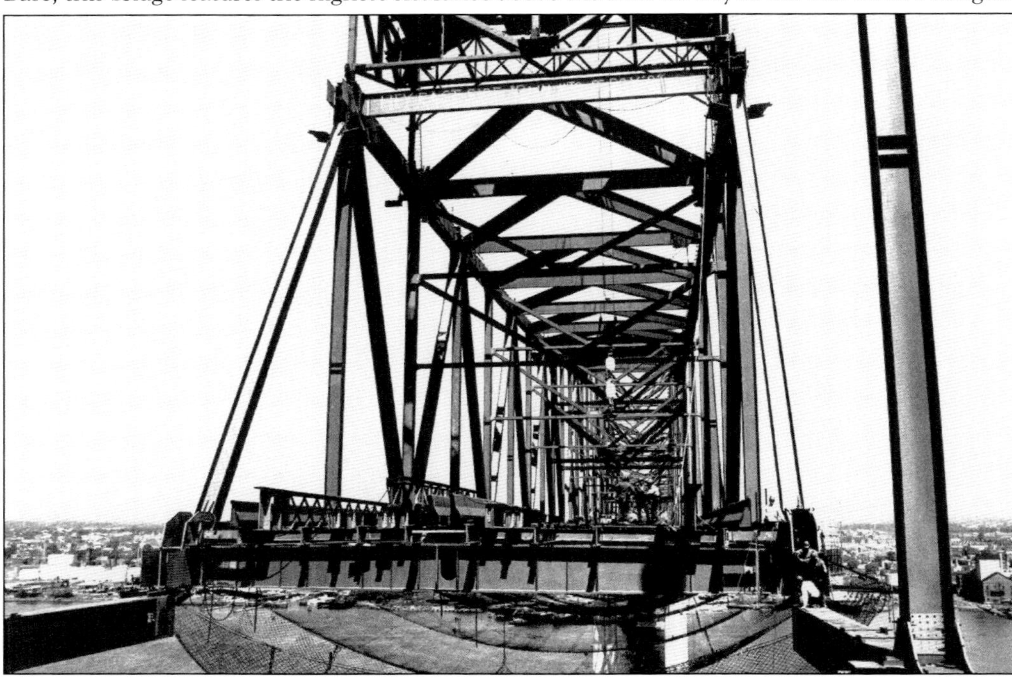

As construction continued on both sides of the river, a friendly rivalry developed between the crews on the east and west ends of the bridge. This rivalry is evidenced by the comments written on the top lateral bracing, above the roadway, reading "Hey East Side 'Run Into It, Boys.'"

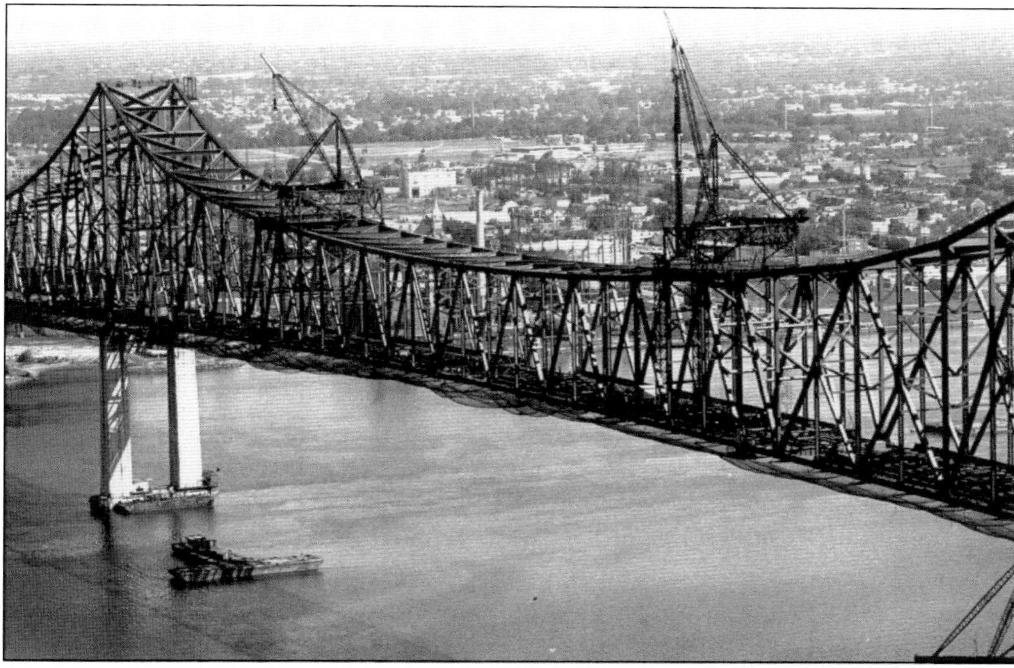

As construction on the Commodore Barry Bridge's steel superstructure is completed, workers are pouring concrete for the roadway surface. When the Commodore Barry Bridge opened in 1974, it sealed the fate of the Chester-Bridgeport ferry service, which can be seen operating to the right of the bridge in the image below.

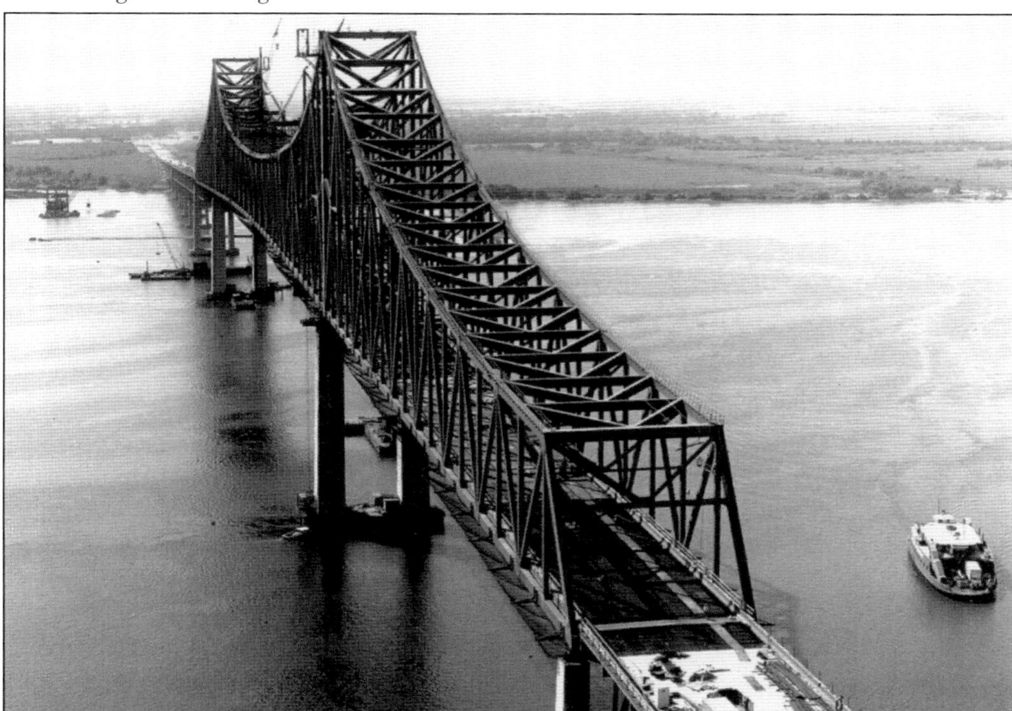

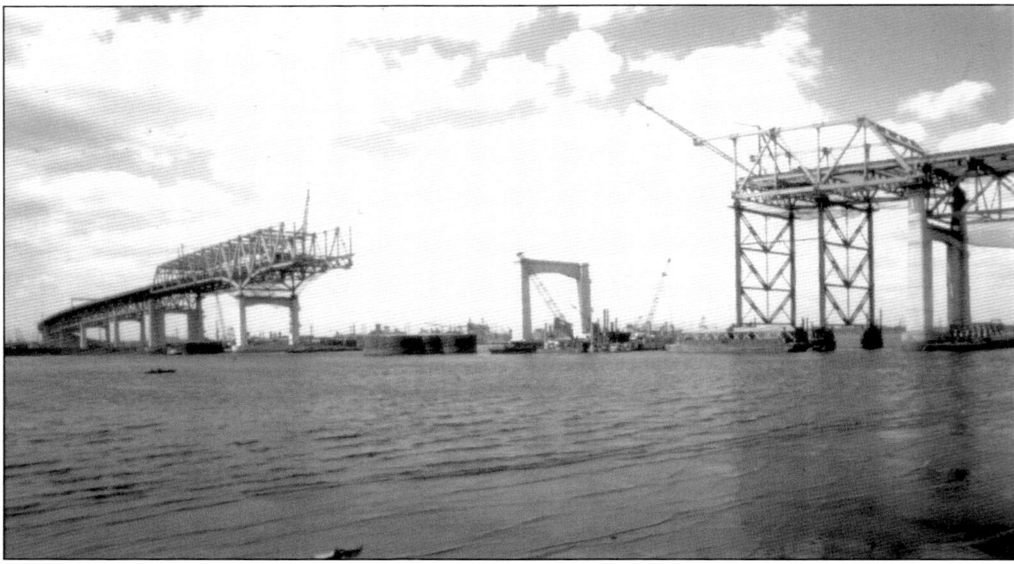

The Betsy Ross Bridge, originally an eight-lane structure connecting Philadelphia and Pennsauken, New Jersey, opened in 1976. The bridge was initially designed to be incorporated into a new highway that paralleled Route 73 in New Jersey and provide a connection to Interstate 95 (Delaware Expressway), U.S. Route 1 (Roosevelt Boulevard), and the southern terminus of Route 309 in Philadelphia; neither highway section was built. For several years after the Betsy Ross Bridge's opening, access on the New Jersey side was limited to U.S. Route 130. The approach roadway was extended by the New Jersey Department of Transportation in the late 1980s, when N.J. Route 90 was extended to N.J. Route 73.

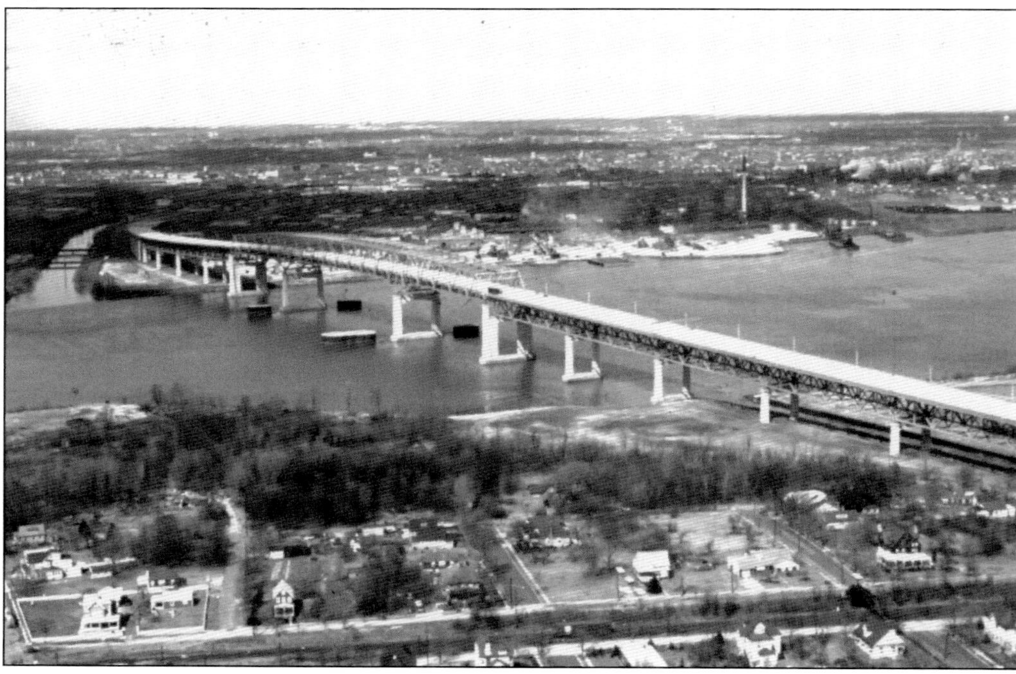

Five

The Bridge Today

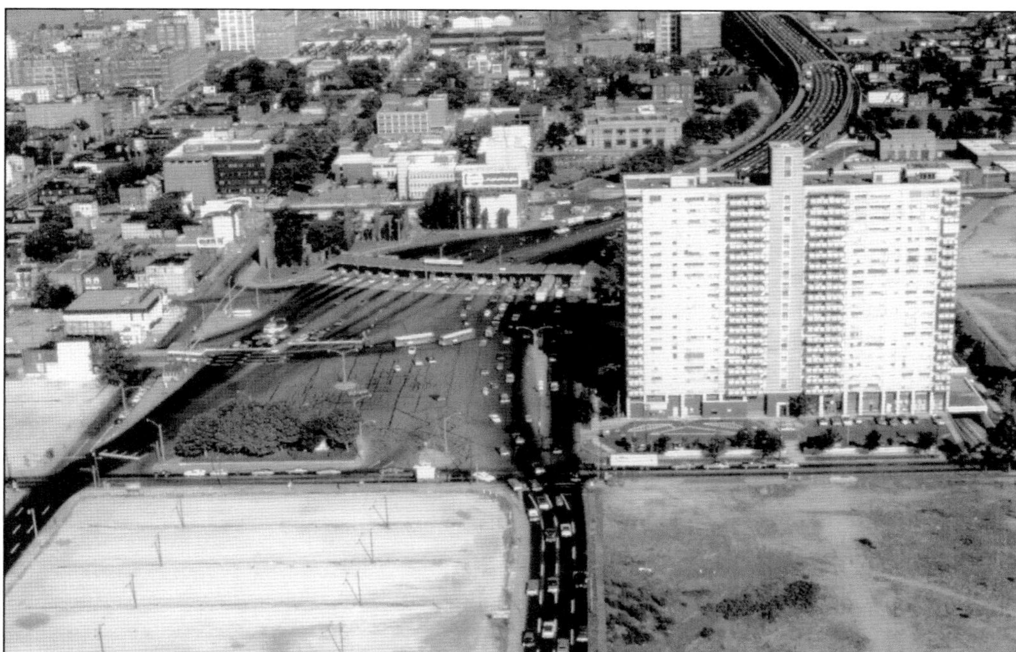

In the 1970s, access to and from the bridge remained the same at the Camden plaza. Although minor improvements were made to improve traffic flow within the plaza, the State of New Jersey would make changes that would completely alter the approach to the bridge. This image shows the plaza conditions prior to the reconstruction.

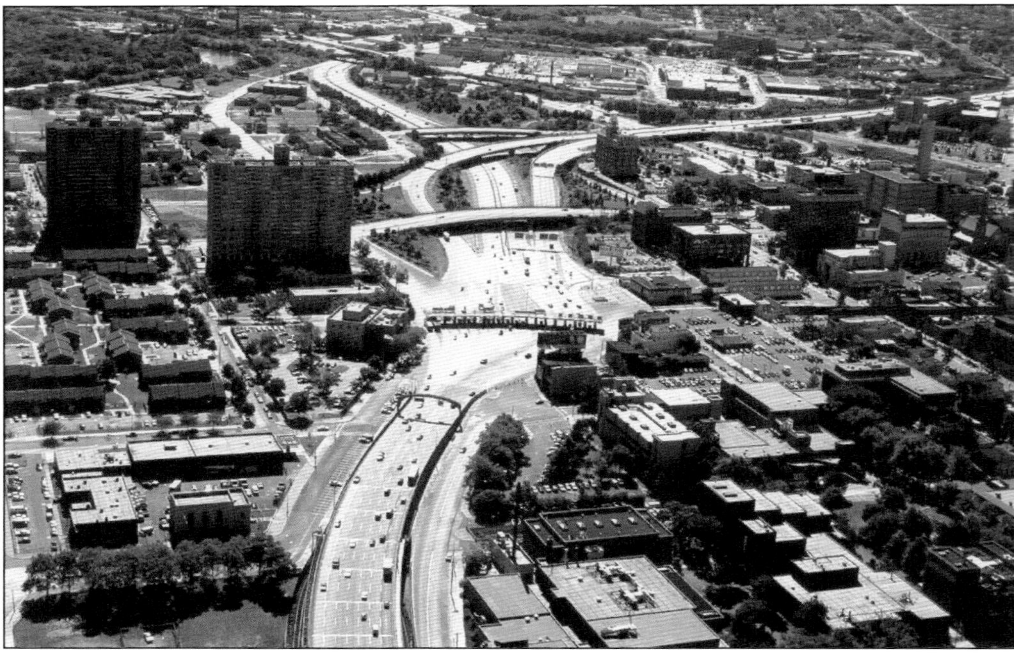

The Camden plaza was completely altered when the State of New Jersey constructed highway connections between the Admiral Wilson Boulevard/Interstate 676 and the toll plaza. As a result of this construction, access to the bridge from Broadway was severed.

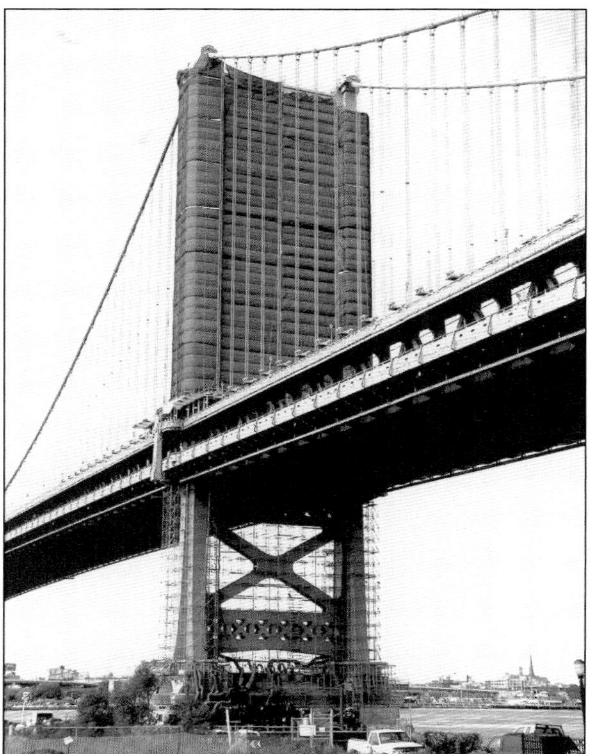

In 2002, the DRPA started a multiyear project to remove all 26 layers of paint from the Benjamin Franklin Bridge to prepare for its repainting. Due to the presence of lead paint, large sections of the bridge were enclosed, such as the Camden tower, as seen here in 2004. This project exceeded the total cost of building the bridge in 1926. Originally gray, the bridge was painted blue in 1976, the year of the nation's bicentennial.

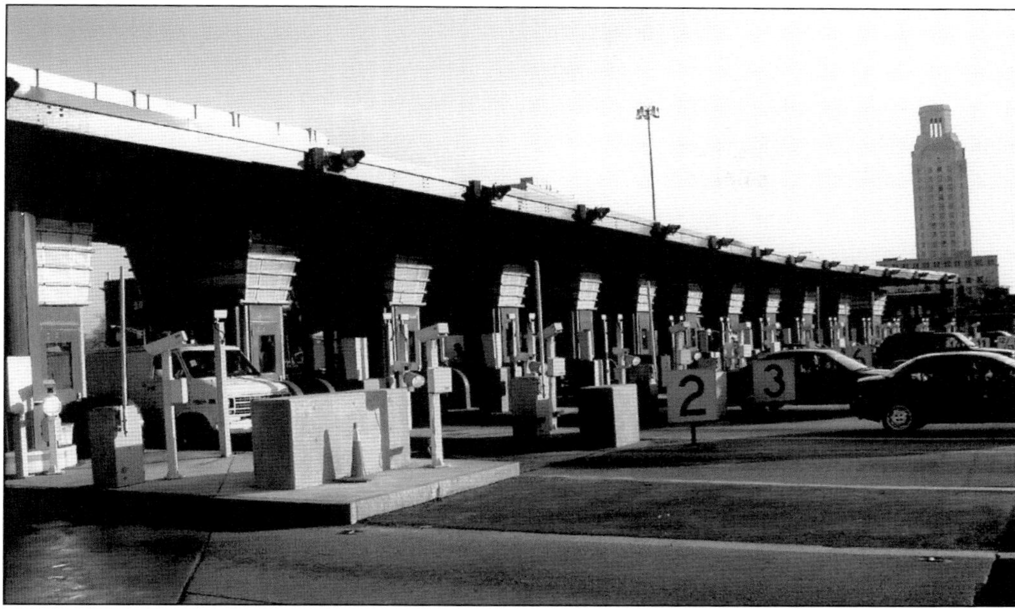

The toll plaza has undergone two improvements to the throughput of traffic, namely one-way toll collection and E-ZPass. In October 1992, toll collection was eliminated in the eastbound direction, and a double toll was collected in the westbound direction. Abandoned tollbooths were removed in the mid-1990s during a toll plaza modernization project. In December 1999, the implementation of E-ZPass allowed tag holders to pass through the toll lanes without stopping.

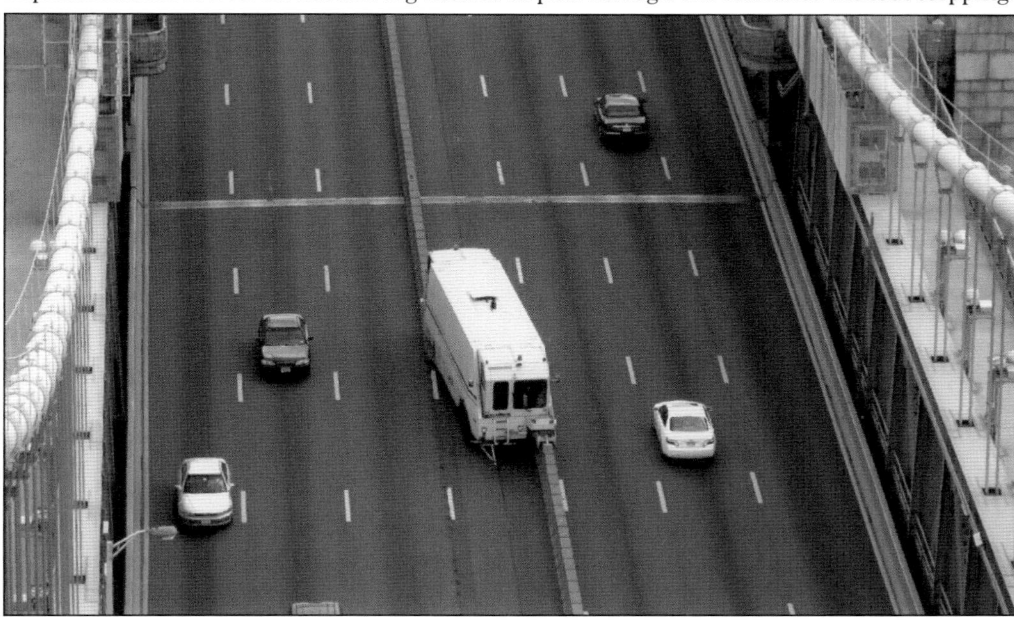

From 2000 to 2001, a "movable barrier" was installed on the Benjamin Franklin, Walt Whitman, and Commodore Barry bridges. The movable barrier allows the DRPA to increase the number of lanes for traffic by utilizing the "buffer" lane, a closed lane between eastbound and westbound traffic. This installation continued the DRPA's history of maximizing traffic flow given the existing physical constraints of the bridge. (Authors' collection.)

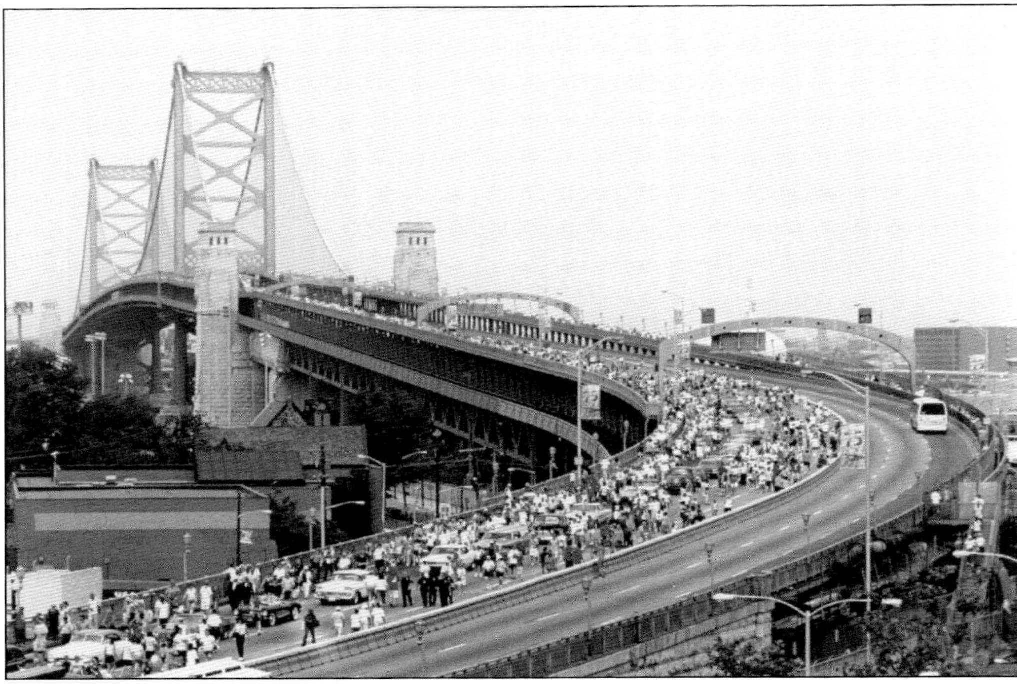

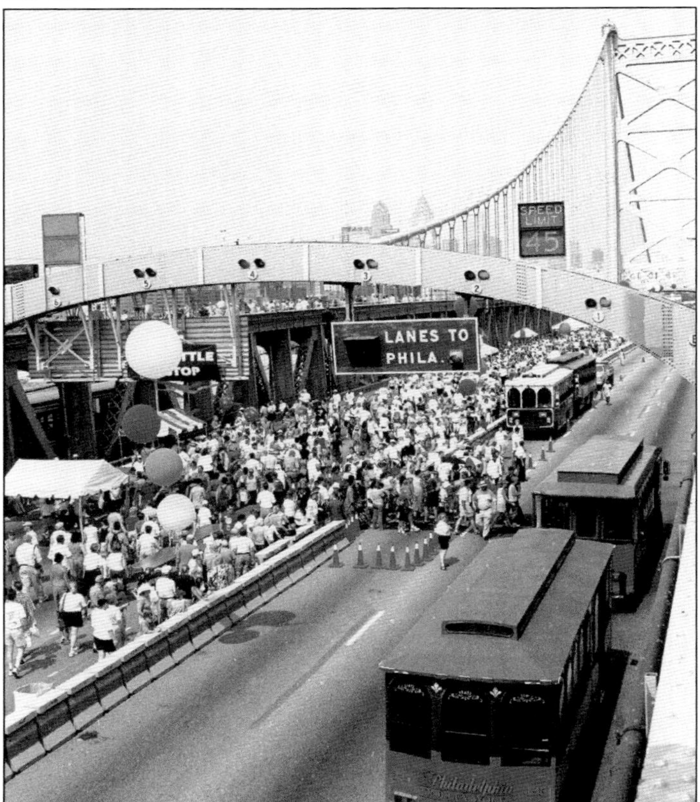

July 1, 2001, marked a celebration commemorating the 75th anniversary of the Benjamin Franklin Bridge. As on its opening day in 1926, for several hours, the bridge was open only to pedestrians. On display for the crowd were antique automobiles, original Bridge Line transit cars, and the refurbished bridge angels. In addition to the ceremonies, which included speeches from local dignitaries, a Benjamin Franklin impersonator, and those who were part of the original opening-day festivities, the Camden anchorage was opened to the public to allow viewing of the decorative tiles in the unused train station lobbies and the cable splay rooms.

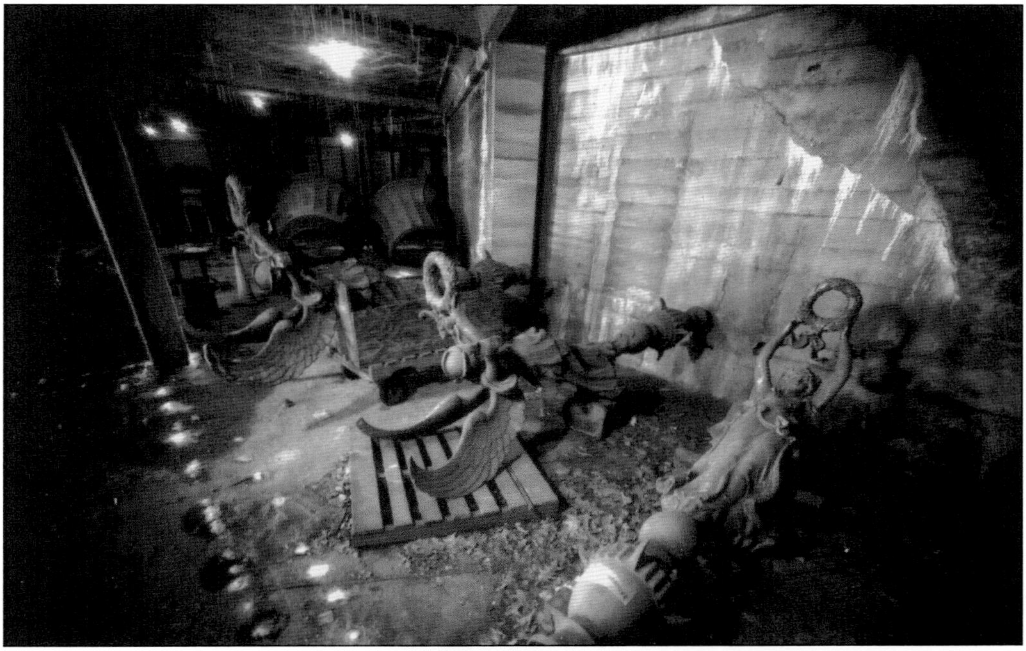

In 1949, the four "Winged Victory" angels were removed from the bridge plazas and placed into storage inside the Philadelphia plaza. The image above shows three of the angels resting on their wings in the plaza storage area. The angels remained in the plaza until 1999, when they were removed and restored. The restoration included cleaning and reapplication of gold leaf to the surface. The restored angels were unveiled during the Benjamin Franklin Bridge's 75th anniversary celebration. The image at right shows one of the four angels currently on display in the lobby of the DRPA office building in Camden. (Above, courtesy of James B. Abbott, © 2000; right, authors' collection.)

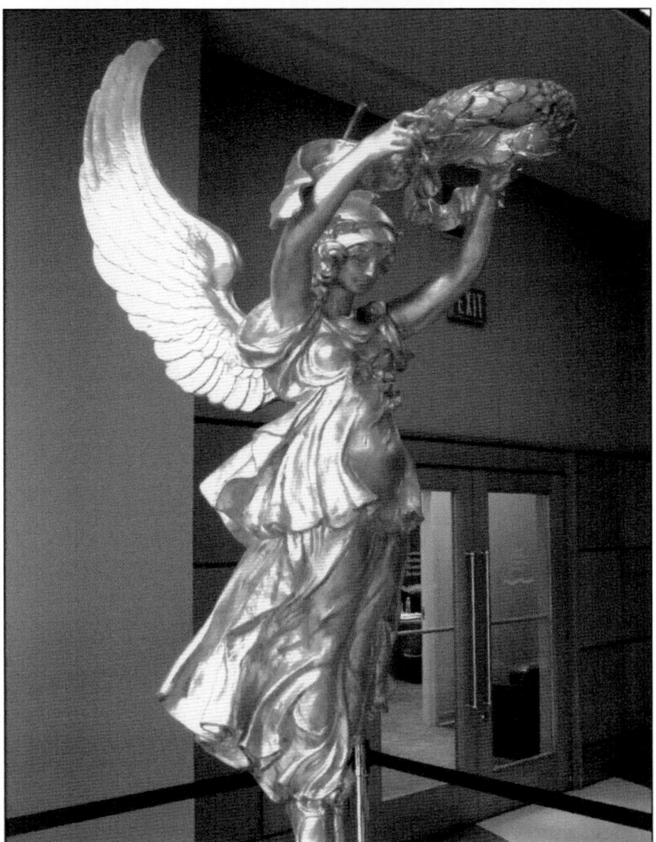

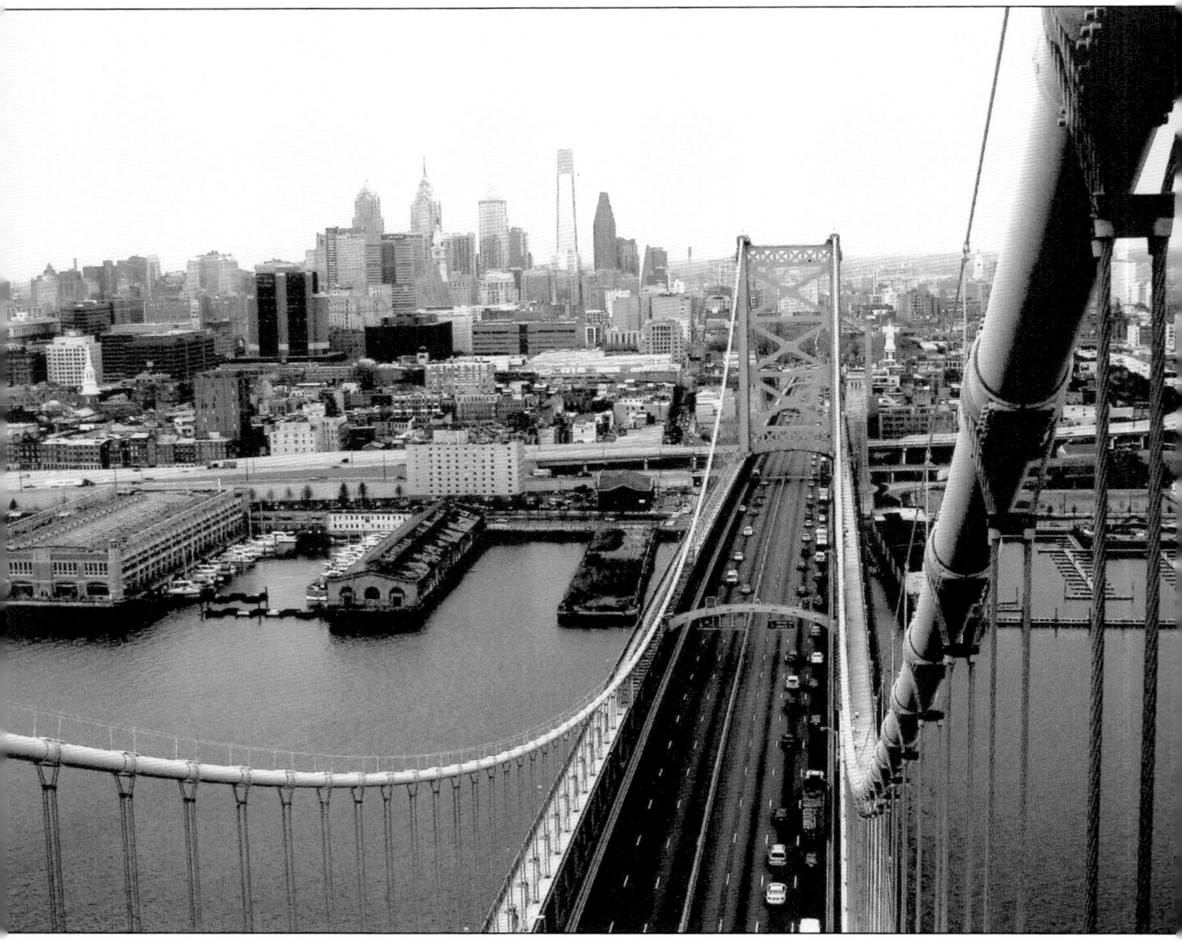

Although many things have changed since the bridge's opening in 1926 and this 2008 image, the Benjamin Franklin Bridge remains an enduring symbol for Philadelphia and the South Jersey region. As DRBJC member Theodore Boettger stated during the anchorage cornerstone laying on December 12, 1924, "We are building here something that will endure and be at its full usefulness after all present here today are gone. These walls of concrete and granite are built to last for centuries. The Engineers tell us that this bridge will serve for hundreds of years if ordinary care can be taken in painting and protecting it. Compared to this structure, man seems puny." As the bridge approaches its 100th anniversary, the DRPA continues its daily maintenance, improvements, and rehabilitations to meet the original engineers' expectations. (Authors' collection.)

BIBLIOGRAPHY

Andariese, W. *History of the Benjamin Franklin Bridge*. Camden: Delaware River Port Authority, 1981.

Beamish, Richard J. "250,000 March Across Bridge After Opening." *The Philadelphia Inquirer*, July 2, 1926, p. 1.

Boyer, Charles S. *The Span of a Century, 1828–1928: 100 Years in the History of Camden as a City*. Camden: Centennial Anniversary Committee, 1928.

Camden, New Jersey: Its Past and Present. Camden: Publicity and Welfare Committee of City Council and the board of trade, 1915.

Carswell, Charles. *The Building of the Delaware River Bridge Connecting Philadelphia, PA and Camden, NJ*. Burlington, New Jersey: Enterprise Publishing Company, 1926.

City of Camden, New Jersey. "Initial townsites." *Camden, New Jersey*. http://www.ci.camden.nj.us/history/townsites.html.

Cooper, Howard M. *Historical Sketch of Camden*. Camden: Horace B. Ketler, 1909.

"Commemorating the Dedication of Camden Bridge." *Camden Courier*, July 5, 1926, Bridge edition.

Delaware River Bridge Joint Commission. *Report of the Delaware River Bridge Joint Commission of Pennsylvania and New Jersey*. Burlington: Enterprise Publishing Company, 1920–1937; 1940–1941; 1947–1950; 1952–1956.

Delaware River Bridge Joint Commission. *The Delaware River Bridge Connecting Philadelphia, PA and Camden, NJ: Final Report to the Board of Engineers to the Delaware River Bridge Joint Commission of the States of Pennsylvania and New Jersey*. Burlington: Enterprise Publishing Company, 1927.

Dorwart, Jeffrey M. *Camden County, New Jersey: The Making of a Metropolitan Community, 1626–2000*. New Brunswick, New Jersey: Rutgers University Press, 2001.

Petroski, Henry. *Engineers of Dreams: Great Bridge Builders and the Spanning of America*. New York: Alfred A. Knopf, 1995.

Ryan, Carolyn J. "Philadelphia." *Funk & Wagnalls New World Encyclopedia*. New York: Funk & Wagnalls, 2002.

Streissguth, Tom. "Introduction: The Roaring Twenties." *The Roaring Twenties*, Eyewitness History, Revised Edition. New York: Facts on File, Inc., 2007. *American History Online*. Facts on File, Inc. http://www.fofweb.com. Accessed March 3, 2008.

Weigley, Russell F., ed. *Philadelphia: A 300–Year History*. New York: W.W. Norton and Company, 1982.

Consistent with our mission to preserve history on a local level, this book was printed in South Carolina on American-made paper and manufactured entirely in the United States. Products carrying the accredited Forest Stewardship Council (FSC) label are printed on 100 percent FSC-certified paper.

MADE IN THE USA